DATE DUE

DEMCO 38-296

Cool

Cool

How Air Conditioning Changed Everything

Salvatore Basile

Fordham University Press | New York 2014

Library of Congress Control Number: 2014941036

Printed in the United States of America

16 15 14   5 4 3 2 1

First edition

To Aunt Catherine, who started this

Contents

Acknowledgments

I learned a fascinating lesson during the writing of this book: If you're not a scientist, then the process of investigating a subject that's even remotely scientific becomes—well, not quite research, but something more along the lines of a detective story. And of course, good detective stories depend on those characters who are far more interesting than the detective.

Thanks to John Schaeffer, Steven Jaffe and Stephen Rapp for listening patiently throughout the initial talking phase of this project, and then for reading and rereading and occasionally arguing and always questioning. My appreciation to Lucy Jaffe for pulling information out of dark corners, and to Susie Boydston-White for performing digital salvage on crumbling old pictures. For their time and generosity in opening doors and sharing unusual pieces of knowledge, thanks (in alphabetical order) to Glen Clugston, Brendan Corrigan, Salvatore Diana, Robert Evers, Deirdre Gagion, Christopher Hyland, Joe McNulty, Joseph Manzella, Andy Oaker, David Skoblow, Salvatore Spataro, and Charles Warren.

In addition, my gratitude to the interest and helpfulness shown by staff members of the Brooklyn Historical Society Library; John Ferry of the Estate of R. Buckminster Fuller; Gino Francesconi of the Carnegie Hall Archives; Todd Gilbert of the New York Transit Museum; Jane Gunderson of the Scharchburg Archives, Kettering University; Michael Homolka and the staff of the Minnesota Historical Society; Chris Hunter of MiSci: The Museum of Innovation and Science; Ron Hutchinson and The Vitaphone Project; Eric Kozubal of the National Renewable Energy Laboratory; Janet Linde of the Archives of the New York Stock Exchange; Ryan McPherson of the Baltimore & Ohio Railroad Museum; Mary Milmoe and Ashley Barrie of the Archives of the Carrier Corporation; Polly Nodine of the Jimmy Carter Presidential Library and Museum; Liz Perez of the Friedrich Air Conditioning Co.; Nilda I. Rivera of the Museum of the City of New York; Fred Schmidt and Jack Sheridan of the Saugatuck-Douglas Historical Society; Elizabeth Scott, Exhibitions Manager of the

Cabinet War Rooms, Imperial War Museums; Arlene Shaner of the New York Academy of Medicine Library; Kajette Solomon of the Bridgeman Art Library; and Giema Tsakuginow of the Philadelphia Museum of Art. Finally, a special thank-you to staff members of the Picture Collection of the Mid-Manhattan Library; they were helpful, knowledgeable, and unfailingly patient when confronted with repeated episodes of well-this-is-okay-but-don't-you-have-anything-better. Much obliged.

Introduction

The first air conditioner I ever thought about was a unit that failed to keep Lois Nettleton cool in a 1961 *Twilight Zone* episode. (I was six years old, I shouldn't have been up that late, and as punishment I had nightmares.) In the episode, the Earth was moving toward the sun and heating up to deadly levels, electricity was being rationed, and to emphasize the single hour each day that Miss Nettleton's air conditioner worked, ribbons were attached to its output grille. When they suddenly drooped, everyone on camera became alarmed, and more moist, and the music took an ominous turn.

It wasn't until the following summer that I encountered—at a safe distance, held back by parental supervision—a real air conditioner in a department store display, a clunky tan machine sitting on a table, surrounded by big boxes, and blowing for all it was worth. This one also had ribbons streaming from it. While I didn't completely understand what an air conditioner *did*, I walked away thinking that all of them were required by law to wear ribbons.

Very soon after that, my Aunt Catherine shocked everyone in the family when she bought *two* air conditioners, one for her bedroom and one for the living room. From the (eavesdropped) reactions of various relatives, I learned that this was flashy behavior indeed, also that Air Conditioners Are for Rich People. But at the next family gathering, even though it was a leaden August night, her living room was perfectly comfortable, and it seemed that no one wanted to escape to the porch. When no adult was watching, I cautiously approached the Amana in the window, held my hand in front of the output grille . . . and felt a cool breeze. Astounding. Terrific. Nevertheless, there was a problem: This machine didn't have ribbons. I hoped Aunt Catherine wouldn't get arrested.

This close encounter opened a whole new world of theoretical summer comfort. *Theoretical*, mind you—I didn't get to experience it in my day-to-day life. My parents made do with a battery of fans (one of them was labeled "Eskimo," which even as a child I thought was a gross exaggeration of its capabilities). My school was a dark red brick oven, cooled only by windows that were never opened far enough. Uncle Jim's Soda Shop, the neighborhood resort for the elementary education set, had a single oscillating fan *putt-putt*ing near the ceiling. The man who called himself Uncle Jim sweated through the summer, trying to sell penny candy that had lost a great deal of its charm after baking every night in his shop. Behind the counter was a poster advertising Frozen Milky Way. With wobbly baby teeth, I tried one. One was enough.

It's no surprise that throughout my formative years I became a connoisseur, or more properly a junkie, of air conditioning. I slowed my pace when passing neighborhood movie houses to feel the rush of cold air pouring out the front doors, and I rejoiced at seeing the various frost-covered signs telling moviegoers IT'S COOL INSIDE. I wondered why Ricky and Lucy Ricardo, and Rob and Laura Petrie, seemed cool and comfortable in any season. I kept a mental list of what businesses were cooled and what businesses weren't, and I made them part of my teenage hang-out plan. In every book I read I automatically noted if the people portrayed were overheated: wondering how Atticus Finch could put up with the temperature and taking particularly vicious satisfaction in *The Great Gatsby* and the fact that it forced its immensely rich characters to swelter through the hottest day of the year with no relief even in a luxury suite at The Plaza. And because I wasn't as rich as Jay Gatsby, when I was outfitting my first apartment I bought the cheapest available box fan and carried it with me from room to room like a puppy.

But finally came the day when I moved into a place that came with its very own . . . *air conditioner.* There was a moment of complete disbelief before I could bring myself to turn it on for the first time. Then bliss.

Fast-forward to the twenty-first century. I live in New York, a city that doesn't exist without air conditioning. On a given summer afternoon, you can stroll down any residential street and be dripped upon by someone's condensation. (Apparently this has been nicknamed "A/C Pee.") Look up; most of the windows are occupied by droning metal posteriors. Surrounding you is an unmistakable, comforting hum that issues from battalions of compressors. Everyone accepts this as the way it has always been.

Not always.

If I ever wondered what urban life had been like when there was no mechanical cooling, I got a first-hand answer to that question on the day my own air conditioner died. At the time I lived on the first floor of an 1850s-vintage building whose side-street location meant that its windows were certifiably useless as a source of air. And the city was in the middle of a heat wave. My air conditioner was replaced within two days. But during those two days, those two unbearable days, I found myself thinking, *So that's what it used to be like—EVERYWHERE?* And I imagined some apocryphal nineteenth-century resident living in the same digs, in the days before air conditioners or even electric fans, and wondering how he dealt with it. Especially at night.

Strolling along the avenues on a bright hot day, I found the most impressive architecture beginning to take on a new and slightly sinister importance. What buildings had been constructed with any thought to hot-weather comfort, and what buildings hadn't? What did people do, in days past, when those buildings became too hot to endure? Or the streets? Or their clothing? Or, for that matter, their own bodies? In that age of so many mechanical miracles, had there been no attempts to do anything about the heat?

As it turned out, there had. Most of those attempts had been flops; a very few had been successful. But successful or not, nearly all of them had been greeted with skepticism. The scientific and business communities, and the public, had been quick to hail such inventions as the telegraph as "the grandest in the annals of the world." But when it came to a contraption that could *cool the air*—not only did many people not understand why it was necessary, but plenty of them scoffed at the notion that such a thing could even exist. Heat was a fact. Heat was a thing that Heaven sent you. In those days, it was the Good Old Summertime. If the daily death reports told a different story, well, that was too bad.

When a machine called the Apparatus for Treating Air was invented, it was only the latest of millions of gadgets that tried to fight the heat. These contraptions had a history stretching back to Biblical times and before, and they enlisted everything from water power to slave power to electric power, ice made from steam and cold air made from deadly chemicals, "zephyrifers," refrigerated beds, "glacier fountains," ventilation amateurs and professional air-sniffers. It wasn't until 1902 that a barely adult engineer named Willis Carrier showed up in the midst of this chaos to introduce his Apparatus. It was no less preposterous than any of the other inventions everyone else had tried. However, this one worked. And it changed the world.

Still, in the case of this particular world-changing invention, that's only half the tale. Plenty of excellent histories already trace the technical development of air conditioning. But equally important, and often overlooked, is the story of how people *reacted to it* in those far-off days:

sometimes with enthusiasm, but other times with ignorance, or disapproval, or even pious condemnation. Reactions like these are not only very entertaining; they're significant. They help to explain why it took years—decades—for the public to trust that machine, let alone give it a place in their lives.

Finally, we have to remember that every great invention is a triumph not only of technology but also of marketing, not to mention some clever PR. And air conditioning is no exception. The whole world got to read about the day the King of Siam supposedly walked into a Carrier office on a hot day, felt the welcome coolness, and right on the spot ordered a complete system for his palace. Or when Thomas Edison was dying but was being kept comfortable in the August heat by an air conditioner (loaned to him by a multimillionaire banker pal). Even though Edison was fading fast and had never been a great friend to inventions that weren't his, news reports pointed out that he had enough strength to get out of bed and examine the machine with "great interest."

We're interested, too, if nowadays our interest is based on other reasons. Air conditioning is a device that continually makes the lists of Greatest Inventions Ever, while at the same time it's fingered as the cause of global disaster. It's been responsible for transforming everything: modern architecture, modern entertainment, the world's food habits, even the way Big Business washes its windows. In the meantime, air conditioning has saved countless lives . . . while causing countless deaths.

We love it. We can't do without it. And it's doomed.

(Or is it?)

1   Ice, Air, and Crowd Poison

It didn't matter that it was a gala performance of a hit show. The weather wasn't going to cooperate.

The newly renovated Madison Square Theatre had opened in February 1880 with the melodrama *Hazel Kirke*. A tearjerker and a particularly slushy one, it somehow had captured the fancy of the theatergoing public and was heading for a record run. But in late May, as the Madison Square's management was busy advertising the play's upcoming one-hundredth performance, the Northeast was gripped by a freak hot spell, with several days of temperatures topping out in the mid-'90s. Slang-crazy New York had only recently come up with a new catchphrase, a remake of an old scientific term, to describe this situation: a "heat wave." Whatever it might be called, it was guaranteed to make that hundredth performance miserable.

During those years, America was realizing that a heat wave was much more unpleasant in cities than in rural areas: the larger the city, the more brick and stone and human bodies, the more hellishly hot it felt. And New York might be the nation's largest, richest, and most cosmopolitan northern city, but in reality its latitude was more nearly equivalent to that of Madrid. As often as not, its summer heat could be ferocious. Natives knew this. Tourists were frequently surprised.

Visitors flocked to see Manhattan's "skyscrapers" (another new catchphrase) and were dismayed to find those looming buildings so badly designed and so closely packed together that breezes were rare and upper floors were like ovens. They trotted up and down Fifth Avenue, gawking at the sumptuous mansions of the wealthy; had they been permitted to enter, they would have found the air inside those homes as hot, heavy, and oppressive as the atmosphere inside any slum tenement. They had heard about the exciting crowds to be found in the cobblestone streets . . . but the stones themselves absorbed the burning temperatures of the daylight hours, remaining hot well after midnight, bathing pedestrians in stifling heat (and treating them as well to a dose of the city's signature aroma: a blend of garbage and horse dung). They might see stylishly dressed Manhattanites out and about. A closer look would reveal those sophisticates to be red-faced, panting, and unsophisticatedly drenched in sweat. It was considered impolite to notice.

In New York or anywhere else, if it was hot outdoors, it was going to be even hotter indoors. Should a trip to the theater be involved, this meant that the performance would be a memorable experience—unhappily memorable, as the most wretched part of a torturous evening. In the nineteenth century, summer heat was a problem that the vast majority of architects hadn't quite learned how to handle; "ventilation" was the fanciest antidote they could offer. And when it came to theaters, ventilation was practically nonexistent. Since Voltaire's time, the average theater had been designed as a pressure-cooker: windowless (to kill outside light and sound) and swathed in velvets and brocades (to impress the audience), a place in which large numbers of people were forced to sit close together for hours, generating their own considerable bodily heat, without so much as a breeze. Of course audience members would become miserably overheated. There seemed to be nothing to do about it.

Overheated theaters had been an ongoing problem through the centuries, and it got worse the higher one sat. The famous English caricaturist Thomas Rowlandson offered his own take on the close-packed, sweltering "gallery gods" who toughed it out at the uppermost reaches of Covent Garden.
(Library of Congress, Prints and Photographs Division)

For that matter, anyone who felt overheated by the crowding could count on being even more overheated by the lighting. In that pre-electric age, a theater might use a thousand gas jets—on the stage, lined up at the footlights, ranged along the auditorium walls and clustered in the central chandelier—each one a small fire, sucking oxygen from the air while replacing it with carbon dioxide, carbon monoxide, even a dash of ammonia. All of those fires effortlessly pushed the temperature to 100 degrees or higher.

Above the chandelier was the single attempt at temperature control: a ventilator, described by one theatrical insider as "a large circular grating, placed in the centre of the roof." Theoretically, the heat of the chandelier would draw the auditorium's hot air out through the ventilator. It was a system that was almost guaranteed not to work. The indoor air became "vitiated," the polite term for an atmosphere that was oxygen-starved, fouled with nauseating gases, unbreathably hot. And a whole evening of vitiated air was a sickening experience. Men who braved the theater under those conditions steeled themselves for a night of piercing headaches, with clothing completely soaked from their woolen underwear all the way out to their coats. Women, crushed into their corsets, tended to faint from the heat. Or perhaps they would slip out to the ladies' lounge in order to vomit in privacy.

Like it or not, suffering from the heat was a part of summer playgoing. The *Courier and Enquirer* had once called it "a two hours seething with four thousand mortal men and women in a huge cauldron of brick and mortar." The humorist Thomas Hood had gone into even more unappetizing detail, describing the genteel horror of sitting through a play in hot weather:

What a scene of general relaxation! The "perspiration of delight" . . . stands upon every forehead; handkerchiefs, as signals of distress, are flagging with their owners in all directions; curls, unwinding into lankness like cotton balls; and collars curling off from the obnoxious glowing cheek, like the leaves of the American sensitive plant. The thin pale gentleman on our right looks cool, but he is only at a white heat. . . . That stout lady's visage in the left-hand box might pass for Aurora's,— intensely rosy,—and a leash of pearls—(are they not?)—escaped, perhaps, from her tiara, are stealing down her brow. . . . [W]e think the simple glare of the lamps might have accounted of themselves for the head-splitting symptoms of the playgoer.[1]

A majority of New York theatrical managers had learned simply to call a halt during the hottest months, closing their shows until cooler weather arrived. But this performance of *Hazel Kirke* was taking place in May, not in July or August. Everyone understood that midsummer theatergoing was out-and-out masochism; theatergoing in the late spring was more of a game of roulette, a bet against the weather. Once the thermometer had started climbing, this audience had lost the bet. On the night of the one-hundredth performance, playgoers entered the doors of the Madison Square Theatre with grim, set expressions, wishing they were anywhere else.

However, as they stepped from the lobby into the auditorium, they were in for a shock. The streets were registering a temperature above 90 degrees—but the temperature inside the building, astonishingly, was an invigorating 70 degrees. One audience member was the English novelist Mary Duffus Hardy, who had equipped herself for the worst with smelling salts as well as a fan. But she discovered that she didn't need them:

> Immediately on entering, we felt as though we had left the hot world to scorch and dry up outside, while we were enjoying a soft summer breeze within. Where did it come from? The house was crowded—there was not standing-room for a broomstick; but the air was as cool and refreshing as though it had blown over a bank of spring violets.[2]

As men stopped mopping their faces and women put away their fans, they looked around in confusion. A theater in summer weather wasn't supposed to be endurable, let alone comfortable. But a note in the program explained it in a single line:

COOLED WITH ICE

It was one of the first times people had experienced a ventilation system that attempted to cool the air *specifically for their comfort*. More to the point, this was one of the first times that such a system really and truly worked. And it was a sensation.

The Madison Square Theatre instantly became a must-see attraction for New York natives as well as tourists. As for *Hazel Kirke*, it was able to ignore the thermometer and run straight through the summer, shattering Broadway attendance records in the process. But the exact reason for the show's popularity escaped no one. One man told a reporter that he was seeing the play for the twelfth time: "Certainly I like the play . . . but between ourselves, we all know it by heart now, and make up family parties to come here and be cool. It is the coolest spot short of Coney Island to spend an evening . . . ."

"Cooling Devices . . . Are Not Attainable"

There had always been a strange mindset when it came to extremes of weather. On the one hand, cold was regarded with deadly seriousness. From the beginning of civilized time, the poorest peasant had tried to keep a fire burning in his hut. Buildings with stoves or central heating, innovations that caught on in the early nineteenth century, blasted forth hot air without restraint; when Charles Dickens visited the United States, he complained bitterly about the intense, dry heat "whose breath would blight the purest air under Heaven." Come winter, the average person's clothing, already heavy, was fortified with an array of still-heavier outer garments. "Draughts" were terrifying things, strongly believed to be carriers of disease, and the slightest threat of "catching a chill" sent most people racing for medical help. An essay of the period illustrated most people's fears in the bleakness of its title: "Winter an Emblem of Death."

Heat, on the other hand, seemed to be merely a nuisance, something to be ignored—or greeted with giddy humor. In August 1822 the *Charleston Mercury* printed an editorial, "Hot Weather":

> It is impossible to write coolly upon any thing, when the thermometer points day after day above 90. . . . The very juxtaposition of the parts of the body is uncomfortable. One would be more easy if he could separate his limbs from his trunk and put them bye, for a time, in the coolest spot he could find. You may observe how this feeling operates by the frequent raising of the arms from the sides, thus giving an opening for the breeze to fan the heated parts . . . .
>
> Fat people seem in this torrid season to melt, and lean ones look as if they would dry up.
>
> The heat of the day is followed by that of the night. Cooling devices and anticalorics that would refrigerate the sleeper and the sleepy are not attainable by human art.[3]

The editorial did what innumerable other writings did and would continue to do for decades: whine about the high temperature while combining stiff-upper-lip fatalism with comic helplessness. And it complained that the miraculous Industrial Revolution had failed to come up with a single machine that could do anything about the heat.

In those early years, there were few options. People living in the country escaped to the outdoors in hopes of a breeze. City dwellers had to adapt. Lower-income people were restricted to the streets, the docks, any available patch of grass, or their roofs. The well-to-do made a point of leaving town. If they were stuck in town, they could retreat to "pleasure gardens," often some of the few green spots in a city, where there was the promise of cooler air. Better still, there was ice cream; at a time when iceboxes were brand-new and extravagant innovations, any refrigerated food was a lavish treat indeed. Even a simple carbonated beverage was a specialty item. Through much of the century, Lynch and Clarke's "soda-water" shop at 25½ Wall Street was a magnet for any person who found himself in New York during the summer. Displayed prominently in front of the store was a large thermometer (in those pre–Weather Bureau days it was one of the few reliable thermometers in the city, trustworthy enough that most city newspapers used it as their go-to source for weather reports). On hot days, the climbing temperature readings served as their own form of advertising; without fail, by noontime there would be a line of parched businessmen snaking down the block, each one impatient for an ice-cooled refresher.

Very little had changed by the time that another anonymous wit, driven frantic by the relentless summer of 1856, wrote nonsensically in the *New York Daily Times*:

> Drop an iceberg into the crater of Popocatapetl, fill up with claret, add one of the West India islands to sweeten and flavor; then hand us the tower at Pisa to suck the liquid through, and you will oblige us considerably. Nothing less than this can cool our cracking throat, we do assure you.
>
> There is a thermometer hanging over the way. We have taken the trouble of suspending it opposite to a picture of the Arctic regions, and keep a boy continually holding an umbrella over its head—still it stands several thousand degrees above white heat. The thin red column of colored alcohol glows like the essence of fire; rays of flame seem to burst from the little globe at the bottom and burn into our brain. Whatever the heat may be, that thermometer is always hotter. It is a demon thermometer, and is doubtless filled with some charmed blood gathered from the veins of an attorney at the Witches' Sabbath. . . . If we were at this moment to be suddenly iced, we feel convinced that it would be possible to run us into a mould, flavor us with vanilla, and put us on a supper

When indoor rooms became too stifling to endure, many city
dwellers—forbidden by law to sleep in any park—headed to their rooftops.
But each hot night was followed by the next day's reports of sleepers who
had rolled off and plunged to their deaths. (New York Public Library,
Mid-Manhattan Picture Collection)

table. Perhaps—ecstatic thought!—perhaps even the loveliest of her sex might take a spoonful of us to cool her beautiful but fevered lips![4]

Whether such asininities were printed in the *Daily Times* or anywhere else, readers didn't seem to find it strange that they often appeared on the same page as other articles which impassively listed each day's heat prostrations and fatalities, arranged in neat columns: "Manhattan and Bronx—Dead. Brooklyn—Dead." For added effect, those articles would carefully note the precise locations at which people had collapsed (in New York, the large stone plaza in front of City Hall was considered a particularly lethal spot; in Philadelphia, the Navy Yard). And many newspapers would publish temperature readings taken in full sunlight as well as in shade in order to provide their readers with the most thrillingly horrific numbers. Some of those articles reported temperatures that ranged up to 135 degrees.

It might seem that the physical dangers of heat just weren't taken seriously. That was part of it, but the real problem was that those dangers weren't at all understood. Newspaper editorials might rail furiously against "abominable dark inner offices" and homes in which "inmates stifle in a stagnant atmosphere"; still, the majority of buildings, public or private, were designed without any real attempt to provide ventilation.

The prime offenders were commercial and public structures. Factories in particular had been built with an eye to profit, not comfort, and rarely with a thought to employee well-being. Windows were there for light rather than for airflow. The buildings themselves were crammed with machinery, driven by steam engines, that threw its own massive heat into the air along with various types of debris. The oppressive temperature (one medical commissioner stood among the workers with a thermometer and recorded 140 degrees) and its accompanying odor were favorite topics of horrified observers: "Factory work is by no means healthy. The rooms are heated to an unnatural degree"; "They labour in an over-heated atmosphere"; "The general atmosphere of the rooms is hot, moist, steamy, and disagreeable"; "What numbers of them are still tethered to their toil, confined in heated rooms, bathed in perspiration . . ." . But the aristocratic commentator Mrs. Frances Trollope was philosophical: "[I]t is difficult to find any factory properly ventilated—free admission of air being injurious to many of the processes carried on in them."

The most public buildings of all, houses of worship, were in their way as badly ventilated as theaters; they prided themselves on their elaborate fields of stained glass, but few of those windows could actually be opened.

They weren't known as "sweatshops" for nothing. (Library of Congress, Prints and Photographs Division)

The Harvard Magazine tried to treat the problem as light comedy in 1863, discussing the famously suffocating atmosphere of the university's Appleton Chapel: "Indeed, the air is almost as much a part of the building as the walls. Perhaps, from its long residence in the Chapel, it has become, in some degree, sacred and inviolable. . . . On Sunday morning, a small pane of glass in each window is thrown open, but it is soon closed. . . ." Actually, the Harvard congregation was lucky. Most church sextons were notorious for their unwillingness to open those windows in response to hot weather. (When questioned about this, one of them snarled, "I'd just have to close 'em again someday!") As a result, churchgoers became so accustomed to the sight of ladies overcome by the heat being carried out during summertime services that the phrase "church faint" was coined to describe it. Everyone attributed the church faint to overzealous religion, or a bid for attention, or extreme flirtatiousness. But no one took it seriously.*

* Not only women were involved in this. One clergyman complained that, swathed in his Sunday vestments and standing throughout a service in the airless chancel, he often felt faint but didn't dare indulge himself. And the *Police Gazette* reported on a service at New York's cornerstone-of-society Grace Church, on a day so hot that one

Private dwellings weren't much better off. Some thoughtful home designs provided houses with cupolas to vent upper stories, porches and verandahs to provide an escape from the heat, or overhanging eaves to shield windows from direct glare; taking this example to the utmost, the southern "dogtrot" sliced a house right down its center with a breeze-catching space that was part passageway, part living room, and wide open from front to back. But such features were absolutely dependent on the vagaries of the wind: no air movement, no cooling. And those features were by no means universal. An alarming number of rooms in buildings of every stripe, everywhere, ranging from small-town slum dwellings to country farmhouses to New York's costly Albany Apartments, and even the Palace of Versailles, had no windows at all. Surprisingly few people were bothered by this. Windows might be opened during the day to admit breezes, but persnickety householders made sure to close them at bedtime, no matter how hot the night. They had spent their lives being told of the dangers of "night air."

If people felt that they were shielded from winter's cold by the amount of clothing they wore, during the summer they might be justified

The "dogtrot" house was generally situated facing north-to-south, the better to scoop up any available air. (Photo: Billy Hathorn)

---

well-dressed gentleman proceeded to make himself comfortable as he sat in his pew by removing his coat, then his waistcoat, then his tie, *then* his trousers—at which point, ushers intervened quickly to escort him out.

in feeling trapped inside a cocoon of cloth. No matter the temperature, women were tightly corseted, multiply-petticoated, long-sleeved, gloved and hatted. They were allowed no concessions other than lighter colors and fabrics—there was even a category of accessories known as Summer Furs. Men had an equally tough time of it. Winter or broiling summer, no sane man would be seen in public without a "sober" dark wool frock coat, as well as waistcoat, hat, and gloves, and no matter how hot the day he was never allowed to remove a stitch of it. Excessive perspiration was seen not as a sign of impending dehydration but as a social gaffe; ruling over all hot-weather activity was the dictum that horses-sweat–men-perspire–ladies-glow. (Gentlemen were taught to use their handkerchiefs without attracting attention. As no one wanted to acknowledge that ladies actually perspired, etiquette dictated that they merely fan themselves, dabbing only when absolutely necessary. A lady who might find herself fading away in public was expected to carry her own smelling salts, most of which were a combination of ammonia-plus-perfume, to jolt herself back from the edge.)

The medical community provided scant help. Around 1850, a theory had been formed to explain the physical distress, as well as the odor, experienced in crowded and overheated buildings—not the elevated temperature, but an unidentified threat called Crowd Poison. While nobody could precisely analyze it, one authority called it "air sewage," and the American Medical Association explained it still more repugnantly as "that morbid influence always generated where men congregate closely, by their living exhalations, or their dead and effete excretions, or both." Crowd Poison quickly became a popular bugbear for the era's medical men, named as the cause of a string of maladies ranging from typhoid to venereal disease. As a solution to the problem, one issue of *The Ohio Medical and Surgical Journal* suggested the universal cure-all, blood-letting.

Even in less drastic cases, plenty of medical advice was based on theory and superstition, and carefully following Doctor's Orders could finish off an overheated person. Woolen and flannel undergarments were sternly recommended for the summertime, and the health profession advised perspiration-drenched people *never* to remedy the situation by removing any clothing: "internal congestion of the abdominal organs and other evils" might result. It was touted that "the best way to endure heat is to drink as little as possible," that "overindulgence in liquids" and "a too free use of cold water" were dangerous practices during hot weather. No surprise that so many people—most of them male—would go through

Dressing for the weather, 1861. Even though it might be August, ladies were required to smother themselves in up to a dozen petticoats. The little boy probably didn't have it much better, as his playtime outfit would be sewn of "sturdy" wool. (New York Public Library, Mid-Manhattan Picture Collection)

a day stubbornly ignoring the heat and then abruptly collapse in the street of sunstroke, often to meet their maker within twenty-four hours.

Not all of these people were completely heedless when it came to the perils of summer. For most of recorded history, inventors had attempted to find ways to fight the heat. Ancient Persians built palaces with "wind towers," smokestack-like structures meant to scoop up faster-moving and

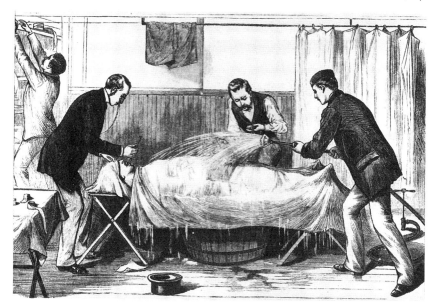

Treatments for heatstroke varied widely. Depending on the doctor's whim, a patient might have leeches applied to the back of his neck, be rubbed with mustard, injected with brandy, given a turpentine enema—or, less imaginatively, doused with cold water until his temperature came down. (New York Public Library, Mid-Manhattan Picture Collection)

presumably cooler air from above the rooftops and funnel it into the interior. Legend had it that a Roman emperor (variously identified as Heliogabalus or Varius Avitis or even Nero) dispatched slave runners to the mountains to bring back baskets of snow, which would then be formed into a man-made mountain in his garden to be enjoyed by him in any way he liked. A second-century Chinese engineer had designed a rotary fan, powered by manual (slave) labor. Leonardo da Vinci, as one of his many duties, had been commissioned to build a machine to ventilate the boudoir of the Duchess Beatrice d'Este; it used a canvas-covered water wheel to scoop up river-cooled air and force it through a series of passageways into her bedchamber. In 1736, the English House of Commons was equipped with a "blowing wheel" seven feet in diameter, hand-cranked by a man designated as, appropriately enough, a "Ventilator." That fan worked only as long as the Ventilator did.

There was also the principle that hot air rises, and this rule was employed by no less a personage than the famed chemist Humphry Davy when he took a stab at the problem in 1811, having been asked to come up with a plan to ventilate the miserably hot and often foul-smelling

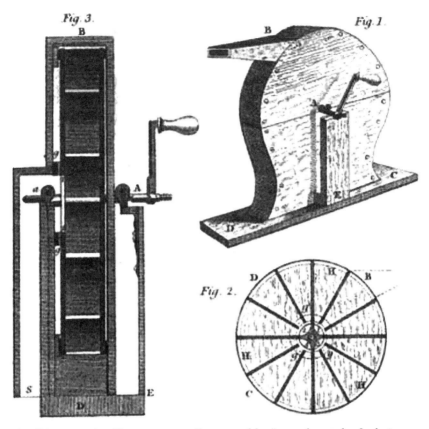

The "blowing wheel" was supposedly reversible, "to suck out the foul air, or throw in the fresh, or to do both at once." (*The Philosophical Transactions of the Royal Society of London,* Vol. VIII)

House of Lords. His solution was the kind that looked completely reasonable on paper: a series of warming stoves beneath the floor to heat incoming air; a pattern of innumerable holes drilled in the chamber floor, through which the heated air would rise; two "apertures made in the ceiling" and connected to funnel-like devices to capture the rising air; and, as the final touch, two stoves above the ceiling, further heating the air entering the funnels and coaxing it out through the roof. Davy theorized that this arrangement would provide a continual refreshing flow of air wafting softly upward through the chamber, all the while carrying away any objectionable odors.

But air currents are unpredictable. One intrepid observer climbed to the roof to check on exactly what was being ejected by the system and was astounded by the result—"The whiff of the gaseous evacuation caused the only indelible impression ever made on our olfactory organs." However, as a cooling system it proved to be an abysmal failure, so much so that the Lords retaliated against Davy by slashing his fee. Gossip-loving Londoners went so far as to circulate a ditty about it:

> For boring 20,000 holes
> The Lords paid nothing
> D—their souls.

Unsuccessful as this example proved, for decades afterward architects built structures that included "thermo-ventilation." Most of these were hybrid versions of the wind tower, gathering prevailing breezes in ducts with extra-large openings and directing them through a building, heating them if the season demanded, and finally exhausting them through a vent that was itself being heated independently. They were iffy when it came to providing actual relief from hot indoor air. And because they were at the mercy of wind speed and direction and needed inordinate amounts of real estate (one English hospital had built a "four foot wide, 70 yard long" intake duct on its grounds), such schemes were unworkable for smaller businesses or private houses. But in their day, they were high technology—and high style; the ethereal and Gothically lacy Central Tower of London's Palace of Westminster is one of the world's most prominent examples of what was really an exhaust stack, part of an 1840s-era attempt to ventilate the Houses of Parliament.

There were other devices, strictly homemade, virtually useless. An English lady visiting America wrote, "I found it a most refreshing practice to place several jugs of iced water in my bedroom during the great heats; the atmosphere became perceptibly cooled." And the housekeeper's guide *Hardships Made Easy* detailed its own remedy for hot weather, which depended more on wishful thinking than anything else:

> In the hot days of summer, especially in houses exposed to the meridian sun, a capacious vessel filled with cold water should be placed in the middle of a room; into this plunge as many green branches as it will hold of lime, birch, or willow tree, with the lower ends in the fluid. By this easy expedient an apartment will be in a short time rendered much cooler, the evaporation of the water producing the desirable effect without any detriment to health.[5]

By the time the British were governing India, a surprising amount of attention was being attracted by several crude devices that were used by natives and governors alike: the *punkah*, the *tatty*, and the *Thermantidote*. The *punkah* was no more than a single-bladed fan, suspended overhead and swung back and forth to move the air in a room. The *tatty* was even more primitive, a screen of woven rush matting hung outside a window and kept moist by periodic dousings. Any entering breezes would theoretically be cooled by evaporation. The *Thermantidote* was an attempt to combine the two ideas: A barrel-like contraption that contained a hand-operated fan, its opening covered with a *tatty*, it would be mounted in a window. A servant stationed outside the building would crank the fan and occasionally pour water on the *tatty*, thus forcing cooled air into the room.

The trouble with all of these systems was that they were passive, relying on lucky breezes; or they required constant tinkering from servants; or, like the "capacious vessel" filled with lime branches, they did nothing whatsoever. Or they had a common lack: They made air move, but they could do almost nothing to lower its temperature. The *Thermantidote* and the *tatty*, meant to cool the air, worked only in dry weather. During the June "Monsoon season," they merely dumped extra moisture into the atmosphere and made unlivable rooms even more so. The *punkah* became the only one of these devices to gain something of a foothold in America, used in military barracks, hospitals, and other public buildings. One writer described the "automaton punka" installed in 1865 at Washington's Lincoln General U.S. Hospital; it was little more than a lengthy rod suspended above the beds in a ward and supporting a row of palmetto fans that waved slowly over each bed. The writer was nevertheless impressed that the arrangement was "capable of being kept in motion by a single attendant, who can thus fan two rows of beds, thirty or more in a row." However, the *punkah*'s manpower requirements limited its use elsewhere; if it was seen in a private home, that home was probably located in the slave-owning southern states.* Various other inventors made attempts to couple it to clockwork- or pendulum-based mechanisms. None of them proved practical.

Perhaps it was just as well. Tennyson had defined the period for all Victorians when he insisted that "self-control" was an essential quality—and this included an ability to ignore, or at least to seem to ignore, discomforts caused by the weather. Moreover, for some people the very idea of a man-made device intended to move the air was nothing less than

---

*The southern version of the *punkah* would achieve Hollywood immortality when it was depicted fanning a roomful of diners in the 1938 film *Jezebel*.

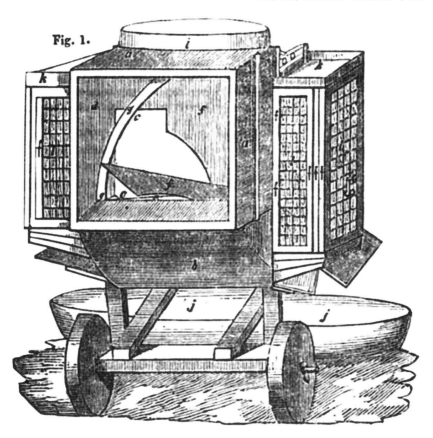

Fig. 1.

The *Thermantidote* was a clumsy machine that could be built by anyone with an assortment of parts. In 1846, *The Mechanics' Magazine* described this specimen in detail, a monster that took two men to operate and was "an acquisition of some value." (*The Mechanics' Magazine, Museum, Register, Journal, and Gazette*, Vol. XLIV)

depraved. Even the original implement to be called a "fanning-machine," an eighteenth-century farm appliance that separated grain from chaff by means of a hand-cranked breeze (and whose fan was nearly identical to that of the *Thermantidote*), had met with stern resistance from many puritans. To point out the mortal error of replacing Divine Providence with such evil contraptions, they quoted the Book of Amos:

> He who forms the mountains,
> creates the wind,
> and reveals his thoughts to man,

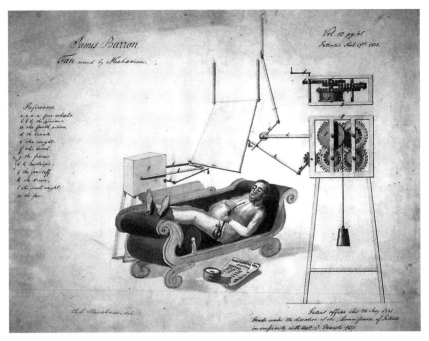

In the 1830s, Virginia inventor (and U.S. Naval Commodore) James Barron patented his mechanically powered *punkah* as a "machine for fanning bed chambers, dining rooms, halls, &c." (National Archives, Records of the Patent and Trademark Office)

> he who turns dawn to darkness,
> and treads the high places of the earth—
> the LORD God Almighty is his name.

Gorrie's Folly

Merely moving the air was better than nothing, but not by much. Cooling it would be better. And even during these years, a few renegades thought that cool air in a hot climate would be not only a relief, but a therapeutic measure. One man in particular became obsessed with the idea.

John Gorrie was a thirty-one-year-old physician when he first hung out his shingle in Apalachicola, Florida, in 1833. It was an odd match. Florida was little more than a wilderness, not yet a full-fledged American state; and Gorrie was a Charleston-raised dandy, an intellectual sophisticate who had been educated in New York and was fascinated by everything around him—he served various terms as the president of a local bank, as an Apalachicola city councilman, as its postmaster, even as its mayor. Some of his neighbors might have found it a mystery that he had chosen to settle in one of the most backward port towns in the South. But Florida was in need of settlers and Apalachicola was seriously in need of medical help, so Gorrie was welcomed with open arms, and his practice took off immediately.

Gorrie was a passionately dedicated medical man, disturbed by the fact that Apalachicola was prey to regular summertime outbreaks of malaria and yellow fever. This was unfortunate but inescapable. The town was bordered on three sides by swampland, swampland was a prime breeding ground for mosquitoes, and mosquitoes were the carriers of those diseases. In the nineteenth century, however, there was no understanding of insect-borne illness. Instead, the medical establishment believed that such diseases were caused by "miasmas," clouds of foul air (very much like Crowd Poison) that supposedly were generated by decaying vegetation in hot, damp places. Whatever the cause, Apalachicola was becoming notorious for scores of fever deaths each year. Gorrie, in an effort to ease his patients' suffering, hit upon the idea that a fever could be completely cured by lowering the patient's body temperature. At first, he created a fever ward in his home in which air moved over buckets of ice suspended from the ceiling; the cooled air sank in the room, settling over the patients in their beds. The patients were chilled, if not cured, but Gorrie had something more in mind.

Students of natural science had known for decades that a gas, compressed, will become hot—and if that gas expands, its temperature will drop sharply. Gorrie took that knowledge a step further, using a small steam engine to power a mechanism of his own design that drew in air, compressed it in a chamber with a piston (becoming hot), and forced it into a labyrinth of pipe. As it escaped into the pipe and expanded (becoming cool), it was routed through a tank of brine, which itself became chilled below freezing and helped to lower the temperature of the air even more. This was already a familiar theory; a number of inventors and physicists around the world, Benjamin Franklin among them, had written on the possible ways in which artificial cold could be produced. However, their writings were fragmentary and theoretical. John Gorrie had actually produced a machine that was complete. More important, it worked. For the first time, a machine was manufacturing cold air.

The modern reader might think that such a device, in a town whose climate was virtually tropical, would have been a godsend. But even as they vigorously plied their palmetto fans against the heat, most of Gorrie's associates simply could not see the value of a machine that cooled the air of a room. Sweating was accepted as a way of life; the very idea

J. GORRIE
ICE MACHINE.

No. 8,080.   Patented May 6, 1851.

Fig:1.

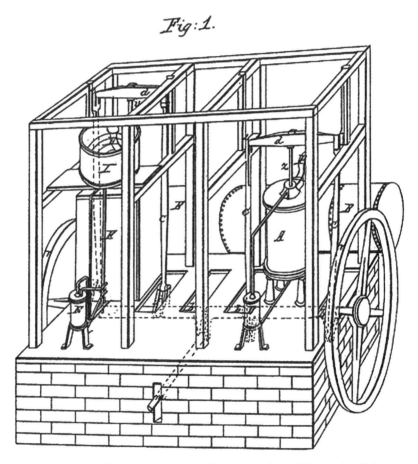

U.S. Patent #8,080: John Gorrie's "Ice Machine." (Records of the
Patent and Trademark Office)

of "comfort cooling" was nonexistent. However, Gorrie believed not only that comfort cooling was a sound theory but that it could be adapted to entire municipal areas—"[W]e are able to cool a city to any degree required by the habits, comfort and health of its inhabitants." In the face of complete skepticism, he wrote enthusiastically of his discoveries in a handful of scientific publications. No one showed much interest at all.

In the meantime, the machine spent some time in his fever ward, running constantly—and as one account went, there came a day when it suddenly stalled. Gorrie inspected the machine, discovering that its pipes had become clogged with frost. This gave him the idea of routing the chilled brine around iron molds that were filled with water. Sure enough, the water froze, producing blocks of ice.

This was a revelation. Cooling the air for human comfort might be seen as mere frippery . . . but the ability to manufacture ice from plain water, whenever it was wanted, opened a whole world of possibilities. Profitable ones.

In the mid–nineteenth century, refrigeration was a new business—iceboxes had been developed only a few decades before—but the ice that cooled them had already become the center of a multimillion-dollar industry, one that hinged on exactly the right kind of weather. There had never before been a way to manufacture ice. It was a completely natural substance, cut in huge blocks from the frozen surface of New England rivers and lakes during the winter, stored for months in gigantic warehouses, and loaded into sawdust-insulated clipper ships to be exported to cities ranging from Boston to Calcutta. With all this effort, it was dubbed "white gold" and priced accordingly as the most luxurious table item, often going for ten times the price of beefsteak. However, its availability was at the mercy of the elements—an overly warm winter would cause a severe shortage, an "ice famine," during the following summer. And shipping was at the mercy of the trade winds, not to mention the caprices of port officials. When a shipment of ice would finally arrive at its destination, often a third of it already melted, its price would be jacked up in a shameless example of the-public-be-damned, to take maximum advantage of its scarcity.

Anyone could see that artificially manufactured ice would sidestep this problem, and Gorrie took advantage of two separate public occasions to demonstrate the usefulness of his machine: once, when the ladies of the local Episcopal church were planning an ice cream social but lacked ice to freeze the cream; and even more publicly, at a Bastille Day banquet held by the French consul, when there was no ice to chill the champagne.

In both of Gorrie's "demonstrations," it was common knowledge that ice had been expected in Apalachicola but the shipment hadn't arrived. Each time, Gorrie amazed the guests and saved the day by providing generous amounts of ice, manufactured by his machine. (The French consul presented his ice on silver platters, ceremonially carried into the dining room by waiters. The church ladies got theirs in large sardine cans.)

Now Gorrie and his contraption were beginning to receive serious attention. His own interest in the device had always been for its therapeutic value rather than its profit possibilities. Nevertheless, enthused by the possibility of a rosy future and encouraged by his friends, he decided to patent the invention.

He applied for the patent in 1850. Almost immediately, his life went downhill in a spiral of gossip and innuendo.

For four decades, nearly all of the country's ice interests had been under the control of Frederic Tudor, an attack dog of a tycoon who had singlehandedly built the ice trade from one tiny ship into a worldwide empire, earning him the nickname of the "Ice King." It was understandable that Tudor would not have welcomed the prospect of his life's work being replaced by a machine-made substitute—especially not a substitute created by an amateur. So, the legend went, as soon as Tudor became aware of Gorrie's machine, he prepared to fight, underhandedly if need be. And while Gorrie had no power and very little money of his own, Tudor had in his arsenal a huge fortune, also fierce determination; his personal motto was I HAVE SO WILLED IT. In addition, Tudor had wide-ranging influence in every area of American business. This influence extended to certain representatives of the press.

What happened next was never directly linked to Tudor. However, Gorrie was stunned to find himself, and his invention, facing a blank wall of universal disbelief. When a man wrote to the editors of *Scientific American*, asking for information on Gorrie's alleged ice-making machine, they sniffed, "We do not know of any feasible plan for producing ice artificially except at an expense so great as to preclude its manufacture for common purposes. If there was any person in our country who could make ice economically, he would not be at a loss where to go make his fortune." The *New-York Daily Globe* labeled Gorrie a "crank" who "thinks he can make ice as good as God Almighty." That derision even showed up on the other side of the world when the *Bombay Times and Journal of Commerce* weighed in, referring to Gorrie's invention as a "cock-and-bull story" and sneering that "the new contrivance is beaten all to sticks by the thermantidote and wet tatties." Public opinion itself was cynical. The

Age of Invention was regarded as a time of miracles. But when one of those miracles seemed to fly in the face of reason—*ice* manufactured with the use of *steam*?—the average man responded with ridicule.*

Gorrie's life became a nightmare. Financial backing was found to perfect and manufacture his machine, but the backer died before anything was finalized. Other backers inexplicably pulled away from any agreement. He began to write letters to money men, then to scientists; the letters went unanswered. After that, he traveled to a number of American cities, trying to interest anyone who would listen. At one point, he was reduced to walking the streets of New York, trying to hawk pamphlets describing the invention. Finally he returned to Florida, utterly defeated, with nothing in hand but his patent. Convinced to the end that Frederic Tudor had destroyed his life—"Moral causes always equally operative with physical, in advancing or retarding the progress of human affairs, and which seems to be generally retarding attendance upon all attempts to unite science and lucre, have been brought into play to prevent its use," he wrote in a New Orleans medical journal—Gorrie died in 1855.

The seventeenth-century scientist/economist William Petty remarked, "I have observed that the generality of men will scarce be hired to make use of new practices. . . . [W]hen a new invention is first propounded, in the beginning every man objects. . . ." This was John Gorrie's fate, the fate of a man who had—purely by chance—laid the groundwork for practical mechanical cooling. He was far ahead of his time, and bitterly knew it; at the same time that he was obliquely incriminating Frederic Tudor, he wrote that his machine "has been found in advance of the wants of the country." In years to come, his achievement would be noted, a museum would be built in Apalachicola to celebrate his work, and his statue would be erected in Washington. But very soon after his death, his name nearly vanished from the public consciousness, virtually unremembered for the next sixty years.

Washington's Hot Air (Part I)
John Gorrie had run into the backward stubbornness of the mid–nineteenth century's commercial world. But more stubborn, and far more

---

*To that end, Savannah's *Looking Glass* told the cautionary tale of a man on an out-of-town visit who "witnessed for the first time the manufacture of artificial ice . . . by steam power." When he reported this to his neighbors back home, he was questioned incredulously and hauled before a church committee, who expelled him for "mendacity . . . because it was 'agin natur' in the first place to make ice with steam."

backward, were politicians of the era. And the most backward and stubborn of all were in Washington, D.C., a soggy and steaming parcel of land in the middle of nowhere that somehow had been named the capital city of the United States. Its climate was almost as brutally hot as that of Gorrie's Apalachicola; its penchant for partisan game-playing was much worse. So, when the subject of ventilation-for-comfort came up, it was seized upon not as a matter of common sense but as a political plaything.

Thomas Jefferson—who had built his own home, Monticello, right at the top of a mountain, cooler air being one of the reasons—had asked that Washington be planned with "light and airy" streets. But that airiness turned out to be impossible; a later writer would describe Washington as "at the bottom of a topographical saucer where moist and motionless air settles with smothering compression." And from its earliest days, the city's 100-plus–degree summers appalled everyone from tourists to Presidents. In 1791, when Jefferson was still Secretary of State, a correspondent wrote to him claiming that "Saltpetre is universally known to be a very powerful refrigerant" and suggesting that if "under the rooms, which are to be erected, for the accommodation of Congress . . . space sufficient were allowed (without descending to the damp) for Magazines of Saltpetre," air would rise up into the chambers "cool & fresh, & purified." Jefferson didn't push that scheme. A few years later, Jeffersonian House member Josiah Quincy would be less helpful but much more exasperated when he wrote, "The heat of the Capitol is noxious and insupportable," insisting that a number of his cohorts had been killed by it.

Over the years this would become a familiar refrain: A member of Congress would collapse during a summer session, or perhaps die, and other lawmakers simply assumed that he had been done in by the heat. As a precaution, pages would routinely pass out ice water and hand fans. Sometimes this didn't help; the Washington climate was blamed for the death of a later President, Zachary Taylor. After laying the cornerstone for the Washington Monument on July 4, 1850—a two-hour ceremony which found him in full sunlight, decked out in a black suit—Taylor became so overheated that he gulped down not only cold water but a whole pitcher of iced milk. This, legend had it, produced such violent bodily shock that it killed him within days.

Things were made hotter by the fact that government buildings were designed for maximum pomp and circumstance, with virtually no thought for the comfort of anyone who had to spend time in them. An early plan of the Capitol had a number of rooms with no windows at all. One version of the House of Representatives was so stifling in summer weather that everyone in town referred to it as "the Oven." A *Washington Globe* editorial claimed that the White House was downright unhealthy "in the latter part of July, the whole of August and September, and October, until there is a heavy frost." But worst of all was the Library of Congress, at the time a triple-height room located in the Capitol, constructed entirely of cast iron down to the doors for fireproofing's sake, and illuminated by eight skylights. The room and its contents would become so frighteningly hot that Librarian of Congress A. R. Spofford told of books' inflicting burns when he picked them up, the leather bindings shriveling into powder as they sat on the shelves.

When the Capitol was enlarged in the 1850s, the adventure of overheated air continued. Plans for the Senate chamber and the House of Representatives called for multiple large windows in each chamber. But Engineer-in-Charge Montgomery C. Meigs didn't like the idea. The human voice would be harder to hear, he declared (this was a sensitive point; the bad acoustics of the old House of Representatives had been notorious for distorting speeches into gibberish), and windows would create wintertime drafts. He insisted that it would be acoustically better and more "healthful" to make those rooms completely windowless, illuminated only by skylights and gas fixtures and ventilated by openings in the skylights—and to provide ventilation, steam-driven fans would pull in air through a large shaft and propel it through floor registers.

Members of both houses, nonscientific most of them and obviously terrorized by the very idea of such rooms, reacted as if they were to be the victims of professional torture. On and off the record, they yelped that the design was "an iron box covered with glass" and "a kind of cellar, where none of God's light or air can come in to them—where they are breathing *artificial* air." On top of it, they were apprehensive that the steam engines could explode.

Absolutely ignoring them, Meigs forged ahead. The original plans were scrapped and both halls were built to his specifications, with "air chambers" beneath the floors, a rotary fan sixteen feet in diameter to draw air into the House chamber—the Senate's fan, designed for the smaller room, was a mere fourteen feet wide—and steam engines in the basement to drive them. The *New York Herald*, which had been enjoying the back-and-forth sniping, called the result "a source of great pride to Captain Meigs, and of great expense to the national treasury."

However, there were problems ahead. Meigs, a man of such personal pride that he had his name stamped into many of the pieces of cast iron that went into the Capitol's structure, had been an officer in the Army

The cast-iron Library of Congress was built to keep out the possibility of fire. The problem was that no one had given much thought to bringing in air. (Library of Congress, Prints and Photographs Division)

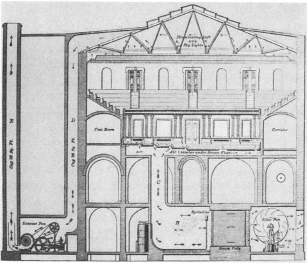

Above: The Senate chamber. *Below:* A cross-section of its unpopular ventilation scheme. Even though Engineer-in-Charge Montgomery C. Meigs had begun by extolling "the purity of the atmosphere of the hall, secured by the perfect ventilation," it turned out that the chamber often topped 90°.
The *American Medical Times* pointed out that "the present system of ventilation in the Chamber was the worst that human ingenuity could devise; the air which [the senators] breathed was pumped up from a damp and unwholesome place below the surface of the ground, and the ceiling was so constructed as to concentrate the rays of the sun upon their heads, giving the Chamber the character of a hothouse for raising exotic plants." (Library of Congress, Prints and Photographs Division/*Knight's American Mechanical Dictionary, Vol. 3*)

Corps of Engineers, not an architect. And while it was Washington lore that he was passionately in love with machinery of all kinds, he was in no way a ventilation expert.

The House chamber was completed first, in late 1857. To test its acoustics, Meigs read—and for extra value Mrs. Meigs sang—from the lectern. However, tests of the ventilation system weren't nearly as comprehensive. Soon after the House had actual people sitting in its seats, everyone realized that it quickly became overheated and, worse, pungently odorous. An observer wrote in disgust, "[N]ot one particle of God's free air can enter it, and it is to be ventilated by an artificial contrivance, like the blower of a steamboat," and the *Herald* chimed in that "whilst it has undergone but little change in respect to ventilation, it has acquired the baking properties of the oven."

Still, Meigs was pleased. Few people had complained, at least to his face, which to him indicated great success. And to illustrate the system's healthfulness, after the House members had spent a season in their new home he proudly pointed out that *not one had died.*

But Meigs, and particularly his reputation, was considerably less lucky when the Senate chamber was occupied in early 1859. Savage criticism began at once, as senators declared that they were "suffering the close heat of an oven" and blamed the "confined and poisonous air of the hall." What happened next was pure Washington: Infuriated references to the ventilation began to pop up regularly in congressional minutes; Meigs— "very much vexed at the unsparing censure heaped upon his superintendence," wrote the *New York Times*—soon found himself assigned elsewhere; and in response to the uproar, a Committee on Ventilation was formed in 1864 to "study the problem."

Almost immediately, the ventilation system began to undergo a costly, embarrassing series of changes.

In a futile attempt to cool the air, a ton of ice was deposited in front of the fans each day. It didn't help. Another try was made when *Scientific American* described "a fine spray or mist, as though made by a gigantic atomizer," discharged into the shaft. Exhaust fans were installed, something Meigs hadn't considered. In 1871 the House floor was entirely ripped out and replaced with a floor that had new and different vents, in an effort to redirect the airflow. And because the air from the original intake was "contaminated by smoke and gas from the neighboring flues," another intake was constructed more than 200 feet from the building. That turned out not to help either, as the new location picked up whiffs of manure that was being spread on the Capitol grounds.

None of these changes did much to improve the temperature of the air, or its smell, and congressional members kept up a barrage of complaints. But that didn't hurt the system's popularity with the general public; the machinery in the basement had been opened to visitors as a tourist attraction, very highly recommended in guidebooks.

Perhaps the ultimate slam was entirely unintentional, showing up in an 1873 *Harper's* article. In great detail, the writer described the halls of Congress, their appointments, their ventilation—and, oddly enough, the floor registers of the Senate chamber. Each register had been equipped with an adjustable lid, the article explained, to give the senators some control over the amount of air they received . . . until it was discovered that they were using them as spittoons.

At that point the lids were fastened almost, but not completely, shut. (Of course, that further restricted the airflow.)*

An Auditorium Cool Enough to Keep Butter Solid

Meigs had been bullheaded—too bad, as there was a germ of usefulness in his ideas. Even in the 1850s, the steam-driven fans he admired were still thought of as exotic hardware (so exotic that the *New-York Tribune* had to describe one to its readers, clumsily, as "two paddles fitted to two rods perpendicular to the axis of rotation") that had first shown up in a few lucky factory spaces and foundries as exhaust equipment. The fact that they made those places more bearable for employees was an unintended, if welcome, surprise. It took a leap of imagination to connect those fans to a system of ductwork, thus creating building-wide air flow, and incidentally the entire business of mechanical ventilation. That business was viewed with awe; but for a long time it would be a remarkably inexact business, in which every step was something of a guess.

One of the very first attempts to use those fans specifically for comfort had taken place a few years before in New York, when the July 22, 1848, *Scientific American* reported, "The Broadway Theatre has a ventilating apparatus in it, which, by means of steam power[,] throws 3000 feet of fresh air into the theatre per minute. By putting an ounce of cologne water into the apparatus, the whole theatre (they say) is made delightfully fragrant—novel certainly."

---

*The nation's capital was a danger even to those who served outside of Congress. In July 1870, Lucien-Anatole Prévost-Paradol, French envoy to the United States, shot himself in the heart. The *New York Times* blamed the Washington climate: "It is presumed that he was laboring under a temporary fit of insanity caused by fatigue and the intense heat."

In those days of infrequent-at-best hygiene, a perfumed theater must have been a treat in any season. But the British engineer David Boswell Reid (the man who had been responsible for that ventilation plan of the Houses of Parliament which included the Central Tower, which got him in hot water with the architect for "interference," after which he was booted from the project outright) elaborated on the basic idea more usefully in the early 1850s when he designed a particularly sophisticated heating, ventilating, and cooling plan for Liverpool's St. George's Hall. Reid's system drew in air, took its temperature, "washed" it by blowing it through fountains, and in the summer blew it over a network of cold-water pipes before routing it into the hall. The cooling component was used for less than fifteen years: The Liverpool water in the pipes was cold, but apparently not cold enough. Nonetheless, the hall would earn a place in some annals as The World's First Air-Conditioned Building.

An ocean away, the same basic idea was being thrown around when New York's plushy Astor House hotel decided to give its businessmen guests a lounging/eating/drinking space by filling in its large courtyard with a $20,000 iron-and-glass rotunda. This was a mixed blessing. At nearly the same time as the rotunda's launch, the city was getting ready for the opening of a much larger iron-and-glass building, the Crystal Palace; it would prove to be a beauteous structure but also a horribly ventilated one, known for mercilessly broiling its visitors in hot sunlight. Perhaps in an effort to avoid this, the *New York Evening Post* reported on the Astor's plans in the spring of 1853, "The immense space is heated in winter by warm air from below; and in the summer, jets of cold air will be blown into the room, moistened by the perpetual play of Croton fountains; and at all seasons the ventilation from the roof will keep the atmosphere fresh and pure." The "jets of cold air" may not have actually existed; while the rotunda lasted for sixty years as a popular restaurant, nothing else was written about its ventilation. Drawings and photos alike would depict the rotunda with windows propped wide open.

After that, a number of Americans ran with the idea, suggesting ways in which fan-driven ventilation might be used along with "refrigeration"—a word that, in those days, invariably referred to ice. These plans were few and far between, and most of them fizzled out before they even made it to the drawing board. When the Academy of Music (a 14th Street precursor to the Metropolitan Opera) was being proposed in 1853, the *New-York Tribune* announced, "[A] system of ventilation . . . proposes the complete exclusion from the building of the external heat in summer by means of double-cased windows, and a perpetual supply of pure, artificially cooled air, which is to be introduced by pipes leading to shafts containing furnaces at the top of the building." There was no explanation of how the air was to be made either pure or cool, and when the Academy was opened in 1854, it had the same ventilation system as most other theaters—none.

Only a year later, the *Daily Dispatch* described an apparatus meant to cool Philadelphia's Walnut Street Theatre:

> It consists of a blower drawing in cold air from the street, refrigerating wheels, and an ice box to cool it, with air tube to properly diffuse it throughout the building. The fan case, refrigerating wheels, box and ice reservoir, are all connected together in one continuous wooden box, the fan being at one end and the ice reservoir at the other. The fan is four feet in diameter by three feet wide, and is driven by a steam engine at the rate of 400 revolutions the minute. The cold air is distributed uniformly by means of eight adjustable openings in the horizontal tubes, and flows into the building at a temperature ranging from 55 to 60 degrees, and so evenly is it distributed that there is no perceptible current coming in contact with any person.[6]

Elaborate as the scheme sounded, it nevertheless was called "primitive"—and it seems to have lasted only a couple of seasons.*

Such systems might have been primitive, but during the early 1860s they were made considerably more practical when a number of inventors from France and England simultaneously announced their "discoveries" of the process for manufacturing ice. Their systems were essentially the same as John Gorrie's (one inventor had visited Gorrie, apparently taking copious notes), but with a significant improvement: Rather than a compressor that used air as its cooling agent, they substituted more volatile substances, particularly ether and ammonia gas. This made these machines more powerful as well as more efficient. As well, the general public—which had once sneered at the very idea of making ice by steam—was now entranced by the concept. The new ice-machine inventors received all the attention that Gorrie had missed; as Europeans, possibly they felt they were out of the reach of Frederic Tudor. Technological

*Exactly a century later in 1955, the director Garson Kanin was in the midst of pre-Broadway rehearsals for *The Diary of Anne Frank* at the Walnut Street Theatre, and he disgustedly noted that the theater had no air cooling at all.

journals and even the popular press began to treat them as modern heroes. Ice slowly began to drop in price and gain in availability, until it would become a household commodity rather than a luxury. As a side benefit, if someone happened to think of cooling a building by ice, such an idea would be considerably more affordable.

Still, the first attempts to use this more-abundant ice were small in scale, crude, and ineffective. One such device was Lesley's Sanitary Room-Cooler, a wardrobe-sized cabinet whose upper portion was given to an ice safe ("just brought out by Mr. Alexander M. Lesley, the well-known inventor of the Zero Refrigerator. . . . [T]he sides of the ice box are perforated, so that the hot air in the upper part of the room may flow over the ice and be cooled. . . . [T]he closet itself always remains cool, and forms an excellent receptacle for articles of food and drink. . . . [A] constant circulation is kept up in a manner exactly the reverse of that which occurs under the action of the ordinary stove used for heating purposes, and the cooling power of the ice is utilized to the utmost"). The principles behind Lesley's contraption might have seemed logical on paper, but with no way to encourage airflow it did almost nothing to cool a room, and before long both Lesley and his Room-Cooler faded from the scene.

If a householder didn't want to cool an entire room, he might spring for a refrigerated bed. Inventor Azel S. Lyman (who had hit pay dirt with a "multi-charge" gun but was responsible for scores of stranger inventions) brought out Lyman's Air Purifier in 1865. *Scientific American* described it admiringly: a hulking cabinet, forming the headboard of a bed, divided into various chambers containing ice to cool the air, unslaked lime to absorb humidity, and charcoal to absorb "minute particles of decomposing animal and vegetable matter" as well as "disgusting gases." Much as with the Room-Cooler, air was supposed to enter the cabinet under its own power, flow over the ice, sink through the lime and the charcoal, and be ejected—directly onto the pillow of the sleeper. "It must continue to blow as long as the law of gravity continues to act." In the same issue of *Scientific American* that described the Air Purifier, Lyman placed a classified ad asking for sales agents. Apparently there weren't many takers.

Once again, a more public demonstration of ice cooling would come from the world of the theater, where the demand was as high as the need. "[T]heatricals do not refrigerate," snarled the *New York World* in a midsummer article. "There is a fortune in store for the man who will furnish us with an Arctic Opera House, an Anarctic [sic] Theatre, the Icelandic Pavilion, or the Freezing Garden." The results seemed to depend

Lyman's Air Purifier, with a rear view of its workings. Air would supposedly enter the cabinet at lower left, travel through the maze, and emerge through slot B "as pure and exhilarating as was ever breathed upon the heights of Oregon." (U.S. National Library of Medicine, Digital Collection)

on the amount of effort that went into the system. Both the Frankfurt Opera House and Vienna's Hofburgtheater used an elaborate arrangement of fans to pull air into "a basement chamber, where in warm weather, sprays of cold water are made to play." Operagoers seemed more or less content with the results, but a visiting musician wrote sarcastically, "The cooling machinery makes so much noise that it can be heard at every soft passage in the music, but then it is so hard worked and *so* suggestive of comfort." On the other hand, New York's prestigious Metropolitan Opera House touted a "cooling apparatus" for whatever attractions might choose to rent it during the hot months; but for its 3,849 audience members, this consisted of a single 12-foot rotary fan that blew air into the auditorium. While its ventilation was admittedly better than anything provided by the Academy of Music, the Met would earn an undignified reputation for intolerable summertime heat.

This problem was in the mind of actor/playwright/producer Steele MacKaye when he took over the lease of the bankrupt Fifth Avenue Theatre in 1879. A true Renaissance man and a Paris-trained actor, MacKaye had been a fixture of the New York theater scene since 1872, achieving a notable measure of success both on and off the stage. And when some private investors gave him the opportunity to build a stock company of his own, in a built-to-order theater, spending whatever he liked and making whatever improvements he wanted, he jumped at the chance.

Over the next year, the Fifth Avenue was completely remodeled. It opened at the beginning of 1880, renamed the Madison Square Theatre, a resplendent 650-seat jewel box of a playhouse that featured such MacKaye-devised innovations as a double-height elevator stage that could make scene shifts in less than a minute, fold-up seats, a Tiffany-designed interior . . . and, for the first time in a New York theater, a well-thought-out air-cooling system. Along with the theater's other attractions, the cooling machinery was touted to the public in a series of newspaper articles starting in February, a drumbeat of publicity that kept up into the summer months and only heightened public interest.

MacKaye's system was built to his own design, and its lavishness—it was billed as costing $10,000—was probably what made it work. Sitting atop the theater's roof, a 50-foot-high intake shaft, topped by a three-foot-wide rotary fan, drew in outside air and passed it through a gigantic bag-shaped cheesecloth filter (theater employees, whose job it was to wash the filter each week, called it "the dust-catcher"). Once the air entered into the ventilation plant, another fan in the basement forced it through a chamber containing racks of ice, then propelled it through a branchwork of smaller pipes totaling more than a mile in length, enabling it to diffuse softly throughout the auditorium. MacKaye swore that the airflow even helped to carry the actors' voices from the stage.

Because of all its innovations the Madison Square Theatre received a carload of press coverage, nearly all of which centered on its cooling machine. MacKaye, ever the showman, conducted numerous tours of the system, explaining it to reporters, members of the public, even other producers. "Most of the experts in the country have examined its intricacies," enthused the Tribune, "and all pronounce it a marvel in its way, and a very decided improvement upon all former methods of ventilating theatres." The Hartford Daily Courant chimed in, "At the Madison Square Theatre the air is cooled by passing it over tons of ice. . . . Mr. MacKaye says that he can make the auditorium cool enough to keep butter solid."

Even so, other theater managers didn't seem at all anxious to burden themselves with the task of ripping up an auditorium just to install a contraption that could supposedly chill butter or audience members. Cooling a theater was an extravagant proposition, not only in equipment cost but also in the number of employees who were needed to run the

The Madison Square Theatre auditorium. There was a built-in ventilator
beneath each seat in the parquet, a feature that most audience members
probably had never encountered. (New York Public Library,
Billy Rose Theatre Division)

machinery. The Madison Square's fans were powered by steam engines,
which had to be stoked; the ice had to be continually replenished as it
melted (theater publicity tried to make a virtue out of the fact that each
performance consumed two to four tons of ice); and these tasks required
not only extra material but also extra manpower. It may not have been a
coincidence that MacKaye had to relinquish financial control of the Madison Square only a year after he had begun.

For that matter, maybe the public wasn't ready for an auditorium cool
enough to keep butter solid. One man wrote an irate letter to MacKaye
after seeing a performance of *Hazel Kirke*: "Many thanks for a severe
cold and sore throat, contracted last eve, in your charming little Theatre."
Stuffed into the envelope, along with the note, was a theater handbill.
The line "Cooled With Ice" was vigorously underscored.

MacKaye would recycle his cooling-machine design a few years later, when he launched the Lyceum Theatre; but during that time, most of his Broadway neighbors refused to follow his example, settling for make-do ventilation systems, or none at all. It took until 1885 for two other Broadway houses—Koster & Bial's, and Wallack's Theatre—to break down and install their own cooling systems, which coincidentally were patterned after MacKaye's. The *Times* wrote that the Wallack's management was thinking of buying its own ice house as part of the arrangement: "Mr. Wallack evidently intends to escape the extortions of the ice companies."

Aside from the world of the theater, the idea of cooling a building hadn't quite caught the fancy of the general public. Part of the problem was that it wasn't an easy thing to get right. A Staten Island hotelkeeper, who had spent some time in the ice cream trade, borrowed from his background when he attempted to cool the hotel's dining room by blowing in air that was chilled by passing it through pipes buried in ice and salt; the procedure was messy and cumbersome, and apparently it didn't last long. As part of Berlin's 1883 Hygiene Exposition, a cooling machine was installed in a restaurant; the machine turned out to be no match for the restaurant's non-insulated construction, which couldn't keep out the summer heat. Other cooling plans were proposed for structures ranging from Vermont dairy farms to the British Museum. These never got any further than the talking stage.

No matter where they were located, the biggest drawback with "cooling machines" was that they sometimes worked and often didn't. As a result, they had earned a terrible reputation. The *Evening World* sniffed that the theatrical cooling machine "is heard of a great deal more than it is felt." And the drama critic of *Music and Art* was obviously enraged by one of those machines when he wrote of a miserable mid-August night at the theater:

> It was at a certain theatre in this city—a veritable Turkish bath. We had been sitting there positively stewing. It was unbearable. The heat seemed to hang in an impenetrable pall over the auditorium. Women gasped, and men ran around and tried to secure ice water. Such an evening as it was! At the end of the second act, unable to endure it any longer, I went out. Flesh and blood is weak when it comes to being baked. As I reached the lobby, and drew a sigh of relief at the prospect of the open air, the manager approached me, and, with a luminous smile, remarked: "Say a good word about our cooling apparatus. It's the best in the city, and keeps the temperature of the theatre several degrees below that of the street."
>
> I do not carry a revolver.[7]

2   The Wondrous Comfort of Ammonia

As the 1880s progressed, the idea of artificial cooling, once mightily scoffed at, was gaining ground—in print, if not in fact. Magazines such as *Scientific American*, *The Technologist*, and *The American Architect and Building News* appealed to the era's technophiles by endlessly glorifying the process of manufacturing ice by machine. The public's imagination was caught, too. Machines themselves, which most people had known only as noisy and dangerous presences in factories, were slowly losing their terrifying air. And even though it was caustic as well as flammable, the ammonia that powered ice-makers (about which the general population knew little, apart from those smelling salts) was being viewed as a magical substance.

The technology became even more popular when builders of ice-making equipment realized they could rig their machinery to generate cold air and thus created the field of mechanical refrigeration. It was initially available only to businesses; one of the first to take the plunge was Brooklyn brewer Samuel Liebmann (whose brewery would go on to find more lasting fame as the producer of Rheingold Beer) in 1870. Other breweries quickly followed suit, as well as meat packers. All of them were delighted with the results. Both industries had been massive consumers of ice for decades, and both of them had been more and more frequently plagued by spoilage. Pollution and sewage dumping were starting to foul even the most pristine New England lakes, which meant that the sparkling, expensive blocks of ice coming from those waters were often loaded with bacteria. As well, there was another filth factor that few people bothered to mention—ice was harvested by horse-drawn cutting equipment; trotting over the ice, horses did what horses will do; and in response, ice companies had never found it necessary to do much more than rinse off their product with formaldehyde. Even though "germs" were a radical concept, plenty of medical men continually warned that ice should be used to cool only containers, not their contents. That advice was routinely ignored by businesses as well as consumers, often with unhappy results. Machine-made cold side-stepped the problem completely.

(Another early customer in the refrigeration market was, logically enough, the Paris Morgue. Officials had tried for decades to preserve bodies with "water and carbolic acid," a method that didn't work at all. In 1878, refrigerating machinery was installed, incoming corpses could be chilled or even frozen solid, and *Scientific American* wrote admiringly that they were completely intact months after arrival: "They have the aspect of marble or wax.")

Now popular journals and daily newspapers had climbed aboard the bandwagon, hypothesizing the benefits of somehow adapting those machines to cool dwellings. Even the staid *New-York Tribune* got in on it, predicting the future of home cooling with surprising accuracy:

> Why should not the civilized man cool his dwellings in summer as well as warm them in winter? In fact, there is no good reason why he should not. Chemistry and mechanics have solved the problem of producing artificial cold. There is no mystery about the process. It is in constant use for

Ice harvesting on the Hudson River in the 1850s. The horses did more than merely walk on it. (New York Public Library, Mid-Manhattan Picture Collection)

the preservation of fruits and meats and in the manufacture of beer. Readers of *The Tribune* will remember an account which appeared a week ago in an article by one of our contributors of the manner in which the vaults of breweries are cooled at a trifling expense by the use of ammonia. The process which produces hoar frost and icicles in these vaults could easily be adapted to lower the temperature of the atmosphere in private houses and hotels to a degree which would make it delightfully cool in the hottest of dog days.

In this matter of artificial refrigeration science is ahead of an intelligent popular demand. We believe the time is not far distant when people will wonder why their ancestors were so stupid as to live without cooling apparatus in their houses. In future generations the endurance of the excessive heat of summer by the inhabitants of great cities will seem as unreasonable as to live through the winter without fires. In that blessed day every house will have its cold air register as well as its water pipes, its furnace and its gas-meter.[1]

These enthusiastic writings were paraphrased and reprinted in newspapers from Roanoke to Abilene. Such enthusiasm might have given rise to the idea that artificial cooling was about to catch on like wildfire. But time and again, the press would announce a plan to "introduce cool air into any business-house or residence"—after which nothing more would be heard. This was logical. The United States was fast becoming one of the richest countries in the world, with one of the greatest concentrations of millionaires, but at that time the wealthiest and most rabid Early Adopter would have no gadget in his home more innovative than a telephone, perhaps a stock ticker. It would have been inconceivable for such a person to envision the *Tribune*'s "cold air register" as a real possibility. And when mechanical refrigeration didn't even exist for household kitchens, it was unlikely that homeowners would imagine that they needed, or even wanted, some sort of elaborate machine to cool all of their other rooms. *The Cosmopolitan* spoke for most people at the time when it wrote, "'Keeping cool' is in many cases a luxury rather than a necessity."

This type of thinking was exactly the sort of obstacle Steele MacKaye had encountered when he had cooled the Madison Square Theatre; a large-scale cooling system might be very enjoyable, but such a system was viewed primarily as an expensive toy. Most of MacKaye's show business competitors would continue to operate as they always had, without any cooling machinery at all.

The same year the Madison Square Theatre opened, conductor Rudolf Aronson decided to make his own bid for summertime business with the brand-new Metropolitan Concert-Hall; rather than any kind of cooling mechanism, it boasted an auditorium ceiling that could slide back on hot nights to expose the evening sky. The roof was designed as a promenade for audience members, who were invited to stroll about the perimeter of the opening, enjoying the view while sounds of the ongoing performance drifted up to them. This was such unusual sport for New Yorkers that visiting the Metropolitan Concert-Hall quickly became a hot-weather craze, listed in trendy publications such as *Godey's Magazine* and *Puck*, which called it "the coolest and the most fashionable place in the city."

Going a step further, Aronson then opened the Casino Theatre, which was crowned by the Casino Roof, the city's first bona fide roof garden. A less entranced visitor might have seen it as no more than a high-priced version of escaping to the tenement roof to seek a breeze, or perhaps a modern attempt to revive the old pleasure garden while shifting it to the top of a building, offering light dining and music along with a few potted

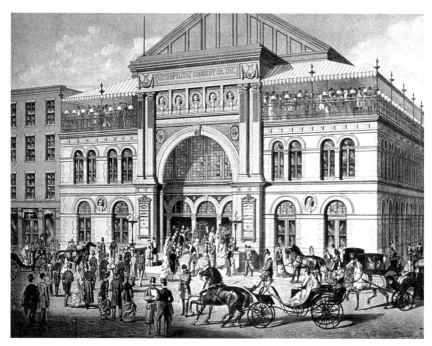

The Metropolitan Concert-Hall, 1880. Note the canopied rooftop promenade, a revelation to genteel (and overheated) music lovers.
(Photo © Barbara Singer/The Bridgeman Art Library)

palms. But everyone considered it a wonderland. The thrill of a skyline view, along with the chance to enjoy some (comparatively) cooler evening air, made the Casino Roof a roaring hit from the moment it opened.

Before long, a number of cities throughout the United States were boasting their own roof gardens atop theaters, hotels, and restaurants. New York had the largest and lushest network of these places, nine of which crowned the city's major theaters, featuring the biggest names in vaudeville and visited every evening by big-spending "roof garden round-ers." Some of them were as elaborate as the buildings that supported them. The roof garden above Madison Square Garden was a super-luxury operation, large enough to accommodate 4,000 patrons. The Paradise Roof Garden wasn't as spacious, but it outdid its competitors in schlock quotient, featuring among its attractions a "village setting" complete with a windmill, a pond, a waterfall, two cows, and a milkmaid. Within a few years, the Hotel Astor would outdo them all by outfitting 20,000 square feet of its immense roof garden as . . . a dirigible station. With an abso-lutely straight face, the manager told the *New York Times* that a garage and repair shop had been installed "so that should an airship party sail up to the hotel they would find ample accommodations for landing."

Roof gardens may have offered fun and escapism, but they had a major drawback—as a refuge from the summer heat, they were useless if it rained. A number of the roof gardens atop Broadway theaters tried to address this problem by erecting open-sided roof structures, which shielded patrons from the rain but only made matters worse on muggy nights by turning the roofs themselves into giant greenhouses. Oscar Hammerstein's Olympia Theatre was equipped with just such a roof, a sixty-five-foot-tall glass canopy; his method of dealing with the problem was to continually pump water over the glass, which may have produced an interesting effect but wouldn't do much about the temperature. Ham-merstein's son Willie, who managed the Victoria Theatre, preferred to manage the situation with deceit. The elevator to the roof garden was *heated* to near-suffocation point, even in August, guaranteeing that as soon as patrons reached the roof they would be gratified by the "coolness" they felt upon escaping.

Even though *Godey's* had been enthusiastic about roof gardens, it unwittingly emphasized the problem of relying on natural ventilation when it wrote, "The audiences that fill the roof gardens of New York are in search of coolness first, and some light entertainment and refreshment second." *Godey's* was right. The annoying truth was that roof gardens

The "greenhouse" variety of roof garden was memorialized by the theater historian Ken Bloom, writing that Florenz Ziegfeld's very first *Follies* was staged in 1907 at precisely such a space—the Jardin de Paris. As he wrote, its glass dome "acted as a huge magnifying lens for the sun's rays. In the days before air conditioning, it was hot; when it rained, the audience held umbrellas over their heads because the dome leaked."
(Byron Company/Museum of the City of New York)

provided the public with architectural interest, titillation, and plenty of spectacle. But they couldn't offer a thing in the way of actual cooling.

Washington's Hot Air (Part II)

Plenty of people never set foot in New York City, never attended the Madison Square Theatre, and knew nothing of the pleasures of roof gardens, but got to read about the ins and outs of mechanical cooling for the very first time as they followed the story of the death of President James A. Garfield.

At the beginning of July 1881, Garfield was traveling to deliver an out-of-town speech when he was shot in the back by a would-be assassin. The wound was serious but not immediately life-threatening; unfortunately, however, the President was then attended by a team of six distinguished (and ambitious, and apparently competitive) physicians. Their treatment—which, in the era before X-rays or the germ theory of disease, included two of the doctors probing the gunshot wound with unwashed fingers in a search for the bullet—guaranteed a slow and lingering demise. And making things even worse for the President was the fact that he had been brought to the White House, which was well known for trapping and holding summer heat. Garfield had been so distressed by his overheated Washington home that he had once wished, publicly, that the place "would only cool off a little."

For the next seventy-nine days, temperatures flirted with the hundred-degree mark, the bedridden President sweated profusely, his condition deteriorated, and the medical team shared their information with the world in humiliatingly complete detail, right down to the daily status of the President's bowels . . . and the fact that he was suffering from Washington's fierce midsummer heat. This news was seized upon by the press and generated a storm of interest from amateurs and professionals alike, who suggested, and sent to the White House, various forms of "cooling apparatus."

At first, the *Boston Globe* reported on an arrangement of flannel sheets "dampened in ice water and hung where the air, passing in through the windows and doors, would become cooled." This invention wasn't a success. Then the *Washington Post* mentioned that "some Boston friends had sent on still another cooling machine"; it turned out to be too small for the size of the Presidential bedroom. There also was a machine designed by a Mr. Dorsey, which was too large (a steam engine large enough to run it couldn't be found, and the thing was dismantled without ever being tried); an oversized icebox, which proved useless because it lacked any air-moving mechanism; and a machine built by a Mr. Jennings that used a small fan to blow air through water-saturated screens. None of them did the trick. As well, there was an idea to adapt one of the city's fire engines "to refrigerate air by alternate compression and expansion." It didn't work out.

After a week, the *New-York Tribune* couldn't help it any longer and began to depict the situation with something of an arched eyebrow:

> In and about the basement there is quite a gathering of engines, boilers, iron, tin and rubber pipes, rope, and other mechanical appliances. A steam-engine stands near the eastern basement door. Upon the greensward and between it and the door is placed a large portable black boiler, with flues, stopcocks and connecting pipes in formidable array. Other

boilers and additional apparatus are on their way. They are to be used for condensing air for cooling purposes which will be expanded afterward in the sick room. . . .[2]

Only moments later, asking one of the doctors about the cooling machines, the reporter wound up getting an earful.

"How about the refrigerating machines?" asked the correspondent.

"Oh, they don't work very satisfactorily. We don't lack for variety, however. It has been suggested that one of them ought to be avoided lest a person approaching it too closely should be overcome by the heat."[3]

And as part of the same interview, Garfield's close friend General David G. Swaim was asked:

"What about the refrigerating machines, General?"

"Oh, don't talk to me about them, or I will feel like throwing you out of the window. I am tired of them."

General Swaim's very evident disgust with the refrigerating machines is probably due to the fact that he has been obliged to experiment with all the contrivances that have been submitted.[4]

But in the meantime, White House officials had contacted Professor Simon Newcomb, a mathematician-astronomer who was attached to the Nautical Almanac Office, assigning him the task of fixing the problem. Although Professor Newcomb was basically unversed in the science of cooling, he felt he could apply mathematics to the situation in order to understand why the parade of machines had failed. He sat down to make calculations, while the President continued to make do with the wet flannel sheets.

As Newcomb was the first person to actually measure the bedroom, it took him no more than a day's work to realize that Mr. Jennings's machine could be modified to give it greater power; Mr. Jennings—Ralph S. by name, a Baltimore inventor responsible for an array of unrelated items ranging from an improved "scarf ring" to a new form of lighthouse illumination—was glad to offer his help. With the addition of a larger engine, a larger "fan blower," and an extremely large ice receptacle, the machine had the ability to lower the temperature of the room to 75 degrees . . . which achievement the President himself confounded by insisting that at least one bedroom window remain open, meaning that the actual temperature of the room would never dip below the mid-80s. In the context of national emergency, no one gave much thought to the fact that the machine's greater power came without great economy, as it consumed a staggering 436 pounds of ice each hour.

Still, it was able to make the President's last days more comfortable—to an extent; by early September a decision was made to move him to naturally cooler quarters at the New Jersey shore. Regrettably, his physicians traveled with him. Still receiving their mostly non-antiseptic care, he developed blood poisoning and died on September 19.

In the wake of Garfield's death, there was national sadness. But there also was a great deal of continuing interest in the "cooling apparatus," enough so that Professor Newcomb was asked to write a comprehensive account. *Reports of Officers of the Navy on Ventilating and Cooling the Executive Mansion During the Illness of President Garfield* was published by the Government Printing Office in early 1882, complete with several pages of tables that demonstrated the machine's operation at full capacity, the data having been compiled *after* the President had been transported to New Jersey and the engineers were finally able to close his bedroom windows.

Later in the year, a Board of Engineers was convened to investigate the device. The Board approved of the results, suggesting that the machine paved the way for similar installations in hospitals and other public buildings but acknowledging the machine's inefficiency: "Had time been offered for experiments, or had experience suggested a more economical method of cooling the President's room, much of the waste of cooling material might have been avoided."

"Waste of cooling material" was a good way to describe it. Two months before that report, an invoice had been received from the Independent Ice Company seeking payment for 535,970 pounds of ice consumed while the machine had been in operation.*

Living Rooms, Refrigerated

If the death of President Garfield and the publicity surrounding all the attempts at air cooling had proved anything, it was probably the fact that ice-based systems were impractical, imprecise, and doomed to be replaced

---

*There was a macabre postscript to this story. The family of Charles Guiteau, the President's assassin, received an offer from a man who manufactured his own brand of refrigerating equipment. He put forth a plan to preserve Guiteau's body after his execution and exhibit it, for a fee, in various American cities. Somehow that scheme didn't materialize.

by artificial cold. *Building Age* nailed it when it declared that "cooling by lumps of ice was crude and wasteful, as well as extremely irregular and uncomfortable."

Over the next decade the image of ice cooling, and even the phrase "cooling machine," took on an old-news tinge. The devices themselves seemed to be demoted to use in theaters, no matter how forward-looking they might be otherwise—Chicago's Auditorium Theatre, which counted Louis Sullivan and Frank Lloyd Wright among its designers and was known for a number of cutting-edge innovations, used a summertime ventilation system that got its nightly chill from fifteen tons of ice. Even Carnegie Hall was designed with a ventilation system that provided a series of ice racks for hot-weather use (although the racks probably were never tried out). Or they were associated with tinkerers' efforts, such as the 1889 Improved Air Cooling Apparatus of a Kansas man that consisted of a screen to fit in a window and a series of "wire ice baskets" suspended in front of it, the whole arrangement designed to drip ice water from top to bottom and, theoretically, cool any air coming into the room. The inventor claimed that his Improved Apparatus was "readily fitted to any sized window or door opening, and easily taken down." It snagged an article in *Scientific American* and was mentioned in a handful of midwestern newspapers, then mercifully it vanished.

Around this time, two events took place that went more or less unnoticed by the general public but were notable for the future of indoor

The Improved Air Cooling Apparatus was an instant fix for any hot room—that is, as soon as a waterproof vat was nailed up on a nearby wall to hold ice and a bucket was positioned below to catch the runoff.
(*Scientific American*)

cooling. First, *Ice and Refrigeration* began publication in 1891, a magazine aimed strictly at professionals and devoted to chronicling the development of refrigeration in all its forms. More than any other journal, it would illustrate the dwindling popularity of the natural-ice trade with the passing of time.

Of greater import, the Colorado Automatic Refrigerating Company invested in large-scale refrigeration equipment, ran two miles of underground ducts throughout Denver's business district, and announced that it was now offering "pipe line refrigeration." This was an innovation, as it offered potential customers the advantages of machine-made cold without the expense or effort of maintaining the machinery. The Denver idea was followed two years later by a couple of firms in St. Louis, one of which claimed that "it will be possible by this plan to regulate the heat in any house from 70° down to zero."

One customer bought into that idea as no one had done before. *Ice and Refrigeration* reported in the fall of 1891:

> . . . the St. Louis Automatic Refrigerating Company has had their system in operation for two or three weeks in their first living room, and their experience so far has shown the system to be an entire success when utilized for this purpose.
>
> The room so cooled is a restaurant and beer hall . . . especially fitted up and decorated to carry out the idea expressed by the proprietor, who calls it the "Ice Palace" on his sign.
>
> The walls have been covered by a scenic artist with representations of incidents in the Kane Polar Expedition, and also with sleighing scenes and other frescoes of a frigid character. The "expansion piping" is placed on the wall about half way between the ceiling and floor, in three sections, or grills. The show window is insulated with double glass, forming the front of a show box which contains the proprietor's name, made up of expansion piping, so that the letters are always covered with a heavy coating of frost, making a very striking appearance from the sidewalk. . . .
>
> No difficulty is found in maintaining any desired temperature in the room, no matter how crowded it may be or how warm the air outside. With the thermometer at 90 deg. to 93 deg. outside, the room is easily kept at from 68 deg. to 70 deg. . . .[5]

Even though the *Tribune* and other such publications had been busy for years predicting just such a use of machine-made cold, the Ice Palace setup was still seen as an oddity—even to the editors of *Ice and Refrigeration*, surprisingly enough, who could find no less awkward way to describe it than "living room refrigeration." In fact, that phrase was used as the title of the article, the magazine's first piece that discussed the cooling of people rather than provisions.

It probably was a good way in which to introduce this machinery to the general public. While the banks of refrigeration coils could have made another restaurant seem like nothing so much as a gigantic meat locker, the Ice Palace's frigid frescoes, as well as its very name, lent a sense of appropriateness and fun. And for most people the frost-covered coils themselves, available for close-up inspection in the dining room and spelling out the name in the front window, made for some fascinating high-tech decoration, vintage 1891.

But no one was fooled into thinking that the Ice Palace installation was a harbinger of the future. It was virtually overlooked in the press, and almost no one seriously considered the notion of adding refrigeration equipment to his home. Average citizens couldn't afford such a thing. Wealthy citizens wouldn't dream of it. The fabulous fortunes of the Gilded Age were being ostentatiously displayed in huge, eye-boggling houses, but the members of the upper class who lived in them (not to mention those wealthy *arrivistes* who were straining to become members of the upper class) were still under the lash of Victorian etiquette. And Victorian etiquette simply would not admit that there was any such thing as unbearable summertime heat. Especially not unbearable heat that required bizarre machines to deal with it.

A year passed before a millionaire—one from California, who possibly felt he had no social standing to lose—decided to opt for a "refrigerated room" in his new home. There were no painted sleighing scenes; in fact, the coils were carefully hidden behind a false wall (which had to be removable, as the gentleman's servants would need to defrost the whole setup on a regular basis). The engine for the machinery had to be placed on the roof, with an "ornamental feature" concealing it. When it was finished, the refrigerated room measured a grand total of six feet by nine. Its cost wasn't made known.

People might be fascinated by the idea of machine-made refrigeration, but it would take far more than a millionaire's whim or a midwestern beer hall to convince them that it was meant for more than food storage. *Scientific American* went so far as to complain about the situation:

> It is a curious example of the slowness with which people take advantage
> of modern inventions that thousands of business men sit sweltering in

hot offices in the midsummer days. . . . At the same time not more than three or four blocks away are great provision warehouses where the temperature is kept at freezing the year round, and at a very moderate cost. If it pays to keep dead ducks and turkeys cool on Greenwich Street, why would it not pay to keep live business men cool on Broadway?[6]

If Not Cold Air, then Moving Air

At the time, less-than-millionaires weren't much concerned with the cost or the clumsiness of machine-made cold. They had fallen in love with another new invention, the electric fan.

A startling precursor had been patented by inventor Louis Stein back in 1854—he had developed a ceiling fan powered *by electricity*: "The rotary motion is imparted by electro-magnets," read the patent application, "connected with a suitable galvanic battery." As no one, anywhere in the world, had electricity, and few people at the time had ever set eyes on a "galvanic battery," Lyman's fan didn't make much headway.

Other ceiling fans, driven by water power and sometimes linked in multiples by a series of belts, had been developed in the 1860s and installed in some public buildings; and there were all of those spring-loaded or hand-operated ancestors of the rotary fan. But these were lazy devices that wafted air slowly. Once Thomas Edison made electric power available to consumers starting in 1882, a host of engineers proposed new machinery to take advantage of it. Within the year, a two-bladed ceiling fan was on sale. Soon after that the first tabletop fans were available, developed by Schuyler Skaats Wheeler, an engineer who had himself logged time in the Edison workshop. They were frightening devices from a personal-safety standpoint, with exposed motor works, two metal blades that revolved at 2,000 rpm ("as rapidly as a buzz saw"; in fact, they were nicknamed "buzz" fans), and no safety grille whatsoever. But consumers—those who could afford the initial exorbitant cost, along with the electric power to operate them, and who had never known any other device capable of producing a fast and steady breeze—thought they were miraculous.

For the last two decades of the nineteenth century the electric fan enjoyed publicity that stood out for its hysterically exaggerated tone, even in that age of ballyhoo. *The Practical Teacher* christened it a "zephyrifer" and asserted that it was ideal for disinfecting sickrooms. The *Columbus Journal* claimed that electric fans were "warranted to lower the temperature of a room from ninety-five to sixty degrees in a few minutes." Businesses ranging from dry goods stores to restaurants and barber shops ran

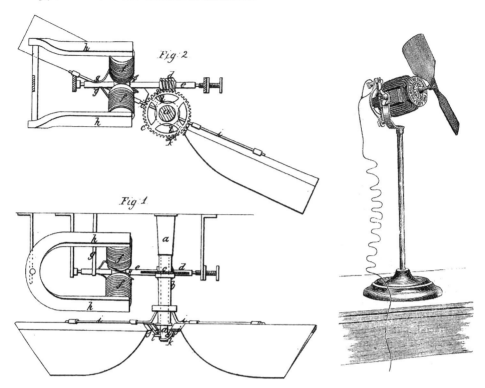

Electric fans: Louis Stein's battery-operated 1854 model (*left*), with its
flapping cloth "wings," and Schuyler Skaats Wheeler's cutting-edge
(literally—note the lack of a safety grille) 1882 version. (Records of
the Patent and Trademark Office and *Scientific American*)

gala advertisements in local papers announcing that their fans would
create "A Cold Wave" . . . "rendering the room cool, cheerful and comfortable" . . . "like going into a cool grove." The *San Francisco Call* tittered
that the fans were showing up even in *churches,* and went even further
to insist, "This is an innovation which will be shocking to many very
devout people, but it seems to us that it is a most sensible thing to do."
And satisfied customers penned joyful letters-to-the-editor, recommending that everyone take advantage of the invention: "This is my third summer's experience with the electric fan, and I don't dread a 'hot wave' as I
used to," wrote a doctor to the *Cincinnati Medical and Dental Journal* in
1887. "I feel so grateful for the instrument that I don't want to keep it to
myself, but want all my friends to know how to keep cool and happy at
the minimum of expense and trouble."

But there were drawbacks. An electric fan was a very high-ticket item, costing up to $20 (more than $475 in modern terms), which restricted its appeal to businesses and the quite-well-to-do. They couldn't be used everywhere, as most areas were still without electricity—and many plants generated power only during evening hours, when lighting would supposedly be needed. And some people were wary of electric fans, rightfully so. At the same time the Cincinnati doctor was praising his fan, the *Lancaster Daily Intelligencer* reported on the buzz fan in A. C. Rahter's hotel bar; the fact that Jake Smith, of the Stevens House Restaurant, thought he could stop it with his little finger; and the result: "The fan still revolves, but Jake's little finger is split from the end to the second joint." (That wasn't merely a small-town accident befalling small-town rubes. The *New-York Tribune* reported a nearly identical story to Jake's; but in Manhattan, it was a broker who received the injury, in a Wall Street banking office. And he sustained cuts on *three* fingers.)*

While Jake's finger healed, other inventors stepped in to offer fans powered by gasoline, kerosene, and even water power. Not many people liked the idea of burning fuel in order to generate a breeze, but the water fan—requiring nothing more than a hose hookup to a faucet, and another one to a drain—was tempting. Also, a water fan could be had for as little as a dollar and a half.

The fact that a water fan was a conservationist's nightmare, with some models using 1,000 gallons of water each day, wasn't considered. But there were other drawbacks. Water fans operated more slowly than electric models, slowly enough that some people found them useless. Just as electric fans were tethered to an outlet, water fans invariably had *two* lengths of hose trailing from them. And hoses could leak. One man wrote to the *Photographic Times* about his unhappy experience when he installed a water fan in his darkroom to dry negatives, the exhaust hose developed a hole while he was away, and he came home to a flood.

If a town's water pressure was lacking, the fan would turn even more slowly. This made them a bad bet for cities like New York, with its famously sluggish water pressure. Manhattanites stubbornly bought the

---

*There was another way in which an electric fan could be used as a weapon. During the 1902 Arkansas gubernatorial contest between Elias W. Rector and incumbent Jeff Davis, Rector dramatically accused Davis of buying "a whirligig fan at the State's expense, to fan himself with, not being content to use a palm leaf like ordinary people." (Even with that kind of scandal, Rector lost.)

fans anyway, with the result that there were summer days during which so many fans were dumping so much water into New York drains that the whole city's water pressure dropped even further, to dismayingly low levels.

"A Great Fountain of Coolness"

"Automatic" fans would take their permanent place as useful tools against hot weather. But whether powered by electricity, water, gasoline, or human effort, they could do only so much and no more when it came to actual cooling. Many people might rely on moving air to get through the summer, but others weren't ready to give up on mechanical refrigeration. A home in Frankfurt took a stab at it in 1892, installing "refrigerated coils" in the attic and allowing gravity to shunt the cooled air through ducts to the rooms below. Virtually no one followed suit; this technology was so expensive, and so incredibly unwieldy, that the mere idea of its being installed in a private house was seen as little more than a stunt.

Artificial cold would have to be demonstrated in portentous, public venues. One of the most public would take place at the 1893 World's Columbian Exposition, which everyone knew as the Chicago World's Fair. For nearly a decade, the Krupp Gun Works had used its own version of artificial refrigeration (using "carbonic acid"—carbon dioxide—instead of ammonia) as part of its manufacturing process. The method was so successful that it wound up with some exhibit space of its own. So the half-million-dollar Krupp Gun Exhibit building featured a strange twist on the concept of indoor cooling, one that received ecstatic coverage in the *Chicago Daily Tribune*:

> A great many people have wished time and again that there was some way of cooling a room in summer in the same way that a stove will warm it in winter. Yet the cooling of the atmosphere of an overheated room seems to be not much of a novelty in Germany. It has been practiced in the great Krupp fabrik in Essen for a year or two. Consequently when Herr Krupp was arranging his great exhibit here he decided, as a matter of course, to also introduce this boon to perspiring humanity. Accordingly, in each end of his great gun pavilion he has erected a great fountain of coolness, which will make it, on many days at least, one of the most comfortable places on the grounds.
>
> These fountains are about twenty feet high, the lower ten feet being of gray artificial stone, treated artistically, like any ordinary fountain, and terminating on top in a circular basin. Rising from the center of this basin

is a spiral of two-inch iron pipe about five feet high. Above that the spiral is about three feet in diameter and five feet higher. The spirals look exactly as if they were designed as a radiator for heating the room by steam: Indeed they look so much that way that no one seems to pay any attention to them or to ask what they are. It is possible also that they are regarded as some sort of mechanical exhibit.

The principle on which these fountains are to be operated is simple. In the engine-room of the pavilion there is to be an apparatus for cooling salt water. It will be exactly similar to the machine for making ice, except that carbonic acid gas will be substituted for ammonia. After the salt water has been cooled down to zero it will be pumped through the coils of pipe in the fountains. At the same time the water of the fountain will be trickled over the coil and frozen to it. This will produce a beautiful frozen cascade, which will by melting cool the air and be frozen constantly by the salt water in the pipe.

It is said that these fountains will cause a delightful coolness in the air of the pavilion. People who are uncomfortably warm will crowd around them and hold out their hands to be cooled, just as they have been accustomed to hold them out to the fire to be warmed.[7]

The *Krupp Exhibition Catalogue* devoted two pages to the "Glacier Fountains," as they were officially named. But they didn't seem to make any impact elsewhere. Visitors and press alike were far more interested in Krupp's star exhibit, "the biggest cannon in the world."*

A few years later the French sculptor-engineer Pierre Roche showed off a still more offbeat use for refrigeration coils. The *American Architect and Building News* reported in 1897, "M. Roche, taking a hint from the fact that pipes carrying liquefied gases become coated with frost, has constructed hollow metal forms of animals, etc. which are chilled by the expansion in their interior of gases liquefied under pressure. A coat of snow, hoar-frost, gathers on their exterior, even in warm rooms, and the snowy figure gives a coolness to the heated atmosphere of balls and feasts."

---

*If fairgoers wanted a more hands-on demonstration of refrigeration's wonders, they'd have to detour to the Midway to try out the Ice Railway, an 850-foot toboggan ride on a track of man-made snow. It operated through the summer months and was such a novel success that, after the fair closed, it was moved to Coney Island; but New York's summer climate was too much for its cooling plant, the whole thing kept melting, and finally its backers gave up.

Less fancifully, refrigeration coils were used to create indoor ice skating rinks. A British invention, the artificially frozen rink had first shown up in America in 1879 as a temporary attraction at Gilmore's (later Madison Square) Garden. By 1896, New York had the St. Nicholas Rink, a deluxe building that offered 10,000 square feet of ice for skaters but was open only from November to March. After a summer or two of costly inactivity, someone had a brainstorm: Keep the rink frozen, install a temporary floor a couple of feet above the ice to create a reservoir of chilled air, use fans to circulate the ice-cooled air throughout the building, rechristen the place the St. Nicholas *Garden*, and present orchestral concerts throughout the summer in cool comfort. The idea was a success; the *Tribune* happily reported that the place was "particularly fortunate in its possibilities of ventilation, a temperature of from ten to fifteen degrees lower than the outer air being maintained by the iceplant."

But none of those ideas would seriously promote the idea of artificial refrigeration for cooling. For that matter, slapping a bank of refrigeration coils up on a wall wouldn't produce real and lasting comfort. Both artificial cooling and ice cooling, as they had been practiced, suffered from a major problem: They did little to address the problem of humidity. Proponents of ice claimed that refrigeration coils did nothing at all about humidity, while "the moisture of the air is condensed . . . and deposited on the ice, to be carried away with the drainage water from the ice; and the air which passes beyond the ice is not only cool, but comparatively dry." Advocates of mechanical refrigeration made exactly the same claims, pro and con. Neither side was completely correct. Both ice and refrigeration coils *could* remove moisture from the air, but there was no way to control the result. Moreover, if the atmosphere was sufficiently humid, either system was capable of producing condensation on a room's walls or, in an extreme case, an actual cloud of fog.

To make matters better—or worse—inventors on both sides of the Atlantic were coming up with devices that attempted to lower humidity. That new technology was embraced by the youthful telephone industry, which was having problems with its "operating rooms." Crammed with heat-generating switchboard equipment and often located on the top floor of a phone company's building, they could become extraordinarily hot on summer days. As well, operators were learning that humidity, dust, and dirt, coming in through open windows, played havoc with delicate phone jacks and their ability to make connections. The Chicago Telephone Company was having a particularly bad time of it, as the central offices were located only a block away from the Chicago River and its smoke-belching

Not merely there for its galvanized beauty, this elephantine "Cold Air Discharge" (as it was admiringly labeled in a magazine article) transformed the St. Nicholas Rink into the St. Nicholas Garden during the hot months. (*The Technical World*)

tugboats. Plagued by up to 400 complaints each day from irate customers, the company decided to take a stab at solving the problem in 1895 with a custom-built ventilation system that would not only cool the air but also clean and dry it—an "air washer." The *Chicago Times-Herald* described what happened as outside air was drawn into the system by a large rotary fan:

> [The air] passes into a tightly-closed chamber, in which a rainstorm is constantly raging. This condition is effected by three rows of small nozzles or atomizers, which discharge a perfect cloud of spray and remove every particle of dirt from the air. The dirt passes off into a well, while the air is whirled through a battery of spiral tubes. The twisting motion, or centrifugal force, removes the last trace of moisture from the air, where it is heated in the winter or cooled with ice in the summer.[8]

While the equipment was installed without a hitch, there was a bad moment when it came time to activate the system; the operators (who at the time were rakishly nicknamed "hello girls") realized that there would be a double-door airlock to the room, and on top of it the windows would be not only closed but caulked shut. Certain that they'd be suffocated, some of them panicked. They relaxed only when the machinery was started and cool air flooded into the room. Delighted, they told the *Times-Herald* reporter that the air was "really as good as if brought from the mountains."

Their happiness didn't last long when they discovered that the system often failed to do its job of drying or cooling the air (the American Society of Heating and Ventilating Engineers itself felt it necessary to warn, "If the humidity of the air, as it enters the washer, is high, then . . . there will be very little cooling effect"). In truth, the system became notorious for the fact that the air outlets mounted in the ceiling would occasionally drip water onto the heads of the hello girls. Other phone companies understandably balked at installing any such device of their own, preferring to make do with electric fans until something more reliable came along. Even so, the Chicago prototype would be patented as the Acme Air Purifying and Cooling System, tirelessly promoted in building journals, a pile of hit-or-miss machinery that would create headaches for architects and engineers for nearly two decades.

Obviously there still wasn't a precise science that applied to indoor cooling, and especially not to humidity control. A great deal of ventilating equipment, and particularly anything that had to do with refrigeration, was still installed by guesswork . . . which could produce some very public failures. When the director of Dresden's Royal Zoological Museum took a tour of American institutions, he noted that the recently built Chicago Public Library had installed an elaborate ventilating system—an Acme—and the building had even been designed with "fixed" windows, designed not to open, to ensure that the cooled air would stay where it belonged. Then came humiliation. Almost as soon as the system was turned on, everyone realized that it was far too underpowered for its space. The nearly new windows had to be expensively torn out and replaced with sashes that could be raised, a number of electric fans were purchased to replace the cooling machinery, and "thus the entire excellently devised system was rendered useless."[9]

Flops of this magnitude might be reported in technical journals, but they almost never made it into the popular press. As the nineteenth century was drawing to a close, writers ranging from H. G. Wells to tabloid journalists were offering a raft of ecstatic predictions, all of them detailing the brave new world that was going to be provided by Science. Along with mile-high skyscrapers, electricity generated by ocean waves, and newspapers delivered by telephone, indoor cooling was a popular subject.

While some of the hoopla had its basis in solid research, most of it was nothing but fantasy. When the development of liquid air was announced in 1898, its 300-degree-below-zero temperature, combined with its portability, seized the national imagination. Magazines as diverse as *The Puritan*, *Current Literature*, *The Dietetic and Hygiene Gazette*, and even *Good Housekeeping* jumped in to predict (preposterously) its use as a cheap and convenient household staple for cooling homes as well as for refrigerating food: "Very possibly, in the near future carts may go from door to door in cities with cans of cold. . . . [I]t may be utilized for cooling the air of the house, a spoonful being deposited here and there in a saucer."

If readers couldn't quite envision Mother bustling about the home with a quart of liquid air and a ladle, they could read other suggestions that had actually made it through the Patent Office but were too ridiculous to provide anything more than comic value. There was the proposal for a gigantic "gas envelope shaped like the section of an orange" that would carry aloft with it a "thick pipe of aluminum, thickly perforated," connected to an extremely long hose that would itself be hooked up to a fire hydrant. Once it was airborne, the water could be turned on, treating the area to a cooling rainstorm. Another idea was provided by an anonymous gentleman, vaguely suggesting a device that would force air through a greenhouse filled with flowers, which somehow would cool the air and scent it as well.

Or topping them all, if only for its blithely imaginative stupidity, was the idea outlined in the *Sun*:

> A Western inventor recently patented a scheme by which he claims he can artificially cool a whole community at little expense. At certain intervals he would erect skeleton towers, like wind-mill towers, each having an electric trolley wire running from bottom to top. The wire transports peculiarly made bombs to a chute at the top, where they are exploded by electricity. The bombs contain liquefied carbonic acid gas, which, when liberated by the explosive, will instantly evaporate and severely chill the surrounding atmosphere.[10]

Washington's Hot Air (Part III)

Back at the halls of Congress, even less had changed. Imitating the Senate's Committee on Ventilation, the House of Representatives formed its own Committee on Ventilation and Acoustics. It was necessary; even after the long series of "improvements" to the ventilation systems, the hot, foul-smelling air that lawmakers received was derided as "unfit to enter human lungs." (The House chamber illustrated this problem more graphically than the Senate. Every morning when the fans were started, a visible cloud of dirt and dust came spewing up out of the floor registers.) As the most recent improvement, and a tacit admission that things were in very bad shape, electric fans showed up in 1890.

A board of advisors—coincidentally, bankrolled by the Committee—had the air analyzed and declared that it was perfectly healthy. One board member went to the expense of traveling to London, to examine the ventilation used by the House of Commons, and brought back the opinion that Washington's system was just fine. To mitigate the odor problem, there was a helpful suggestion that obviously unwashed people should be barred from entering the galleries.

No one in Congress was amused. After nearly forty years of sodden heat, the Senate acted first, appropriating funds in 1895 to fix the problem (part of a final-day-of-the-term marathon session that ran until 5:00 a.m. and ended with reporters in the press gallery spontaneously launching into "Praise God from Whom All Blessings Flow"). To come up with something that would actually work, they appointed MIT professor and engineering superstar Samuel H. Woodbridge to design the ventilation plan.

Although there was general agreement that the problem was "one of exceeding difficulty," Professor Woodbridge devised a series of drastic renovations that involved ripping out the chamber floor and laying an entirely new network of air ducts beneath it; furniture in which "the supports can be made hollow and used as pipes to conduct air"; and electricity replacing steam power and gaslight, a very welcome change as each room had more than a thousand gas jets and temperatures taken at skylight level frequently topped 140 degrees. Finally, as the ultimate touch, *cooling* the chamber "by means of refrigerated air."

As to the cost, various press sources gave out estimates ranging from $4,000 to $15,000.

The work began in mid-1896 as soon as Congress recessed and was nearly completed by December. A few days before the senators reconvened, the system was tested with the fans turned on full blast; because the chamber's carpeting hadn't yet been tacked down, it promptly billowed up and knocked over several seamstresses who were in the process of stitching it. Aside from that mishap, nationwide news reports expressed pleasure with the installation even if they disagreed about the details. *Engineering News* reported that the air "is first cooled to 60° by an ammonia refrigerating apparatus, depositing its moisture, and is then dried and heated to 70° or 75° by passing over hot water pipes," while *The Electrical Engineer* wrote, "An ice plant is to be installed later and means for cooling the air will be provided for sessions in warm weather."

Those publications were wrong when they described the cooling machinery. There wasn't any.

Without mentioning the specifics, everything written about the Senate installation would conveniently omit the refrigeration aspect. But nothing official was mentioned about it until the following year, when the *Report Rendered on the Material Furnished for and Labor Performed in Improving the Ventilation of the Senate Wing of the United States Capitol Under the Appropriation Made Therefor June, 1897* was published. In it, Professor Woodbridge wrote that the project had actually come in far over budget at $55,000, leaving him without enough money to install any refrigerating equipment at all. At the end of his report, he pointed out that "to complete the installation" another $9,000 would be needed. Senate members were so stunned by the project cost that the additional funds never came through.

Even so, Professor Woodbridge was invited to return in 1901 to install a ventilation plant for the House of Representatives, a larger version of the system working in the Senate chamber . . . and this time, he knew better than to try to include refrigeration in the budget. The systems were alike, except for one detail: At the beginning of the work, the *Hartford Courant* reported, the House floor had been ripped out yet once more. And this exposed a major cause of those morning clouds of dirt: "The members have for years used the brass ventilators set in the wainscoting of the hall for cuspidors until the air chambers below have been coated several inches deep with tobacco." To make sure this wouldn't happen again, the air chamber was lined with white bathroom tile and flooded with electric light, and the *Washington Post* assured readers that it would be scrubbed with disinfectant each week. As a topper, it was announced that the chamber would always be available for inspection by any House member who wanted to crawl into the space and have a look.

While that took care of the housekeeping aspect, something had to be done about House members who couldn't control where they spat. So, the *Post* continued, the floor registers were redesigned to eliminate "every opening of a horizontal character, so that it will be impossible . . . for people in the habit of chewing tobacco to *accidentally* [emphasis added] expectorate therein."

Newly ventilated, and at last tobacco-proofed, the House was ready for use in December, and the national press approved: The *New York Times* wrote that "it is now possible to sit for hours in the chamber, crowded though it may be, and not suffer through the effects of a vitiated atmosphere. The merits of the new ventilating system, it is asserted by those responsible for it, will be most apparent in hot weather."

As it turned out, though, the *Times* jumped the gun. Within a few years, dissatisfied House members would be right back to complaining about inadequate ventilation—and discussing the possibility of knocking out a wall of the chamber to get themselves some windows that would open.

Should All Else Fail, Try Science

In a way, Professor Woodbridge had avoided embarrassment when he was unable to install his cooling plant in the Senate chamber; the equipment he had in mind would have been too undersized to do any real good. But he realized that only *after* the fact.

This proved that even the most eminent authorities occasionally worked by blind instinct when it came to the business of ventilation and cooling. But it didn't have to be that way. There was a whole body of scientific research that could make it reliable. The problem had been that a number of people in the business ignored it.

Alfred R. Wolff believed in science.

Wolff was born to the engineering trade, entering the Stevens Institute of Technology at age thirteen and graduating among the top of his class, distinguishing himself with a thesis good enough that it was published in book form.* In 1880, all of twenty-one years old, he set himself up in business as a "steam engineer."

He spent the next decade gathering experience, designing not only heating but also ventilation systems and keeping track of new developments. As so many American systems had proven to be ill-matched to their buildings, Wolff scrapped the calculation tables that nearly every other American engineer used and that were openly called "crude"; in their place he substituted a set of German calculations that more carefully matched the equipment to the space that it was to ventilate. He became known as an innovator, designing installations that featured such cutting-edge technology as cloth air filters, thermostats, and automatic humidifiers to combat dry winter air. By the 1890s it was clear that Alfred Wolff had become the heating-and-ventilation man to the rich and famous (if home refrigeration wasn't welcomed by the upper crust, well-designed furnaces were), with a client list that included such names as the Astoria Hotel, the Century Club, the New York Life Building, St. Patrick's Cathedral, the Brooklyn Institute, the Astor, Carnegie, and Vanderbilt homes—and Carnegie Hall, with its unused ice racks.

While Wolff had been responsible for the Carnegie ventilation system, he plainly hadn't been enthusiastic about ice as a serious method of cooling. There was much more potential in an 1899 commission from Cornell University, which was building its Medical College in New York and wanted not only a ventilation system for the entire building but also a refrigerating system for its fifth-floor dissecting rooms. The Medical College directors weren't particularly thinking of the comfort of the fifth floor's *live* occupants, as they had asked only for a cooling plant that would keep cadavers in optimum condition. Wolff gave them that and far more. His system not only provided reliable refrigeration for the benefit of the cadavers but made students and faculty happy as well; it lowered summertime humidity more successfully than nearly any other installation in existence. (His method, as sketchily described in *Popular Science*, was to "chill the air more than enough"—at which point the moisture condensed on the cooling coils—"and then warm it a little.") It worked

---

*The book, *The Wind Mill as a Prime Mover*, would later be used as a working reference by the Wright brothers as they built their first aircraft.

so well, in fact, that the Medical College started holding its graduation exercises *in* the dissecting rooms, minus the cadavers. Wolff would attend the graduations, hovering in the background and taking careful readings of the air temperatures, outside and in, to chart the system's performance.

While the installation was such a success that Cornell's own *Cornell University: A History* made a point of mentioning it ("These rooms can be cooled by the refrigerating plant in such a manner as to permit the pursuit of practical anatomy with as much comfort in summer as in winter"), its relief from the heat had been, strictly speaking, a coincidence. Wolff still hadn't been asked to install a system intended purely for comfort.

That changed in 1901, when he was asked to design the heating and ventilating plan for the new home of the New York Stock Exchange. Befitting its status as a world money center, the trading floor was conceived as one of the largest interior spaces in the city, a block-long room that would hold 1,500 traders, with a skylighted ceiling 72 feet high and an entire wall of windows that the *Times* supposed would "insure the greatest possible amount of light and air." The building would make an impressive architectural statement. But as any New Yorker knew, in the summer heat it would grill its occupants.

That July, as blasting was still going on to clear the building site, the country endured a murderous heat wave that claimed an estimated 9,500 lives; more than 700 people died in New York alone, where temperatures hit 100 degrees for two consecutive days. On July 4, a surprising interview was printed in the *New-York Tribune*—not by Wolff or anyone connected with the Stock Exchange but by Robert Ogden Doremus. A 77-year-old chemist who held professorships at a number of universities, he still found time to publish articles on a whole range of scientific matters, one of which was artificial cooling. Sitting in his City College laboratory and waving a palm fan, Doremus cited health benefits as well as comfort: "Does anyone doubt that citizens would be happier, merchants more prosperous and physicians able to save more lives with the thermometer at 70 degrees instead of 90 or 100?" He continued, "The new Stock Exchange building, for instance, could be cooled in summer for one-tenth the cost of heating it in winter." Then he drove home the point by issuing a direct challenge: "If they can cool dead hogs in Chicago, why not live 'bulls and bears' in the New York Stock Exchange?"

If this line sounded like a paraphrase of that long-ago *Scientific American* article, it was. But *Scientific American* had stolen it from Doremus himself—he had made the identical "bulls and bears" statement some eight years before in the *North American Review*. Back then, no one had

noticed. Now, whether its message was helped along by the dire temperatures or the *Tribune*'s large readership, the statement rocketed around
New York. Within days it had made its way to the nation's heartland. The
Chicago Daily Tribune even ran a splashy article built on Doremus's ideas,
fancifully illustrating it with a scene of a young executive ordering his
office boy to "*Turn on the cold!*" Coincidence or not, after all this publicity
it became known that the ventilation plans for the Exchange's trading
floor were now going to include *cooling.*

"*Turn on the cold!*"

(*Chicago Daily Tribune*)

Wolff was enthusiastic about the idea, writing to the architect George
B. Post, "This room will be superior in atmospheric conditions to anything that exists elsewhere. It will mark a new era in the comforts of
habitation." At the very least, the project was something of a risk; the
room held more than a million cubic feet of air, and cooling such a cav-
ernous space hadn't been attempted before. But Post—a pioneer in
skyscraper architecture, and the first architect to install elevators in an
office building—was perfectly comfortable with the notion of being a
trailblazer.

There were only the Stock Exchange officials to convince. Wolff cited
economy: As the building was planning to install its own electric plant,

he pointed out that the entire system could be run from the plant's exhaust steam, which meant that the cooling process would in a sense cost nothing, and even the Exchange luncheon room would be cooled along with the trading floor. And there was the comfort factor: Along with cooling, the machinery would be able to wring 2,000 pounds of water from the air each hour. Wolff emphasized this by quipping, "Instead of sticking to your traders' shirt collars, the moisture will run down the drain."

The officials gave him the go-ahead. Even as they did, one of them flatly remarked that if the system failed, Wolff had better buy a one-way ticket out of New York.

The new Exchange opened on April 22, 1903, a day of spring-like weather that didn't tax the capacity of the new ventilation plant. Yet it received news coverage across the country, the first cooling system to make nationwide headlines. The *Evening World* dubbed the Exchange "New York's Coolest Place" in a headline, then continued, "In summer the coolest spot in the city outside of a cold storage warehouse will be the Stock Exchange, which will be kept at a low temperature"—a line that was instantly copied verbatim by papers across the country, including the *Washington Post*, the *Minneapolis Journal*, and even the *Norfolk* (Nebraska) *Weekly News-Journal*. The *Saint Paul Globe* was less hyperbolic, reporting only that "the heating, cooling and ventilating arrangements are perfect." As conservative elder statesman, the *New-York Tribune* merely assured its readers that the installation was "especially elaborate" and reveled in figures when it pointed out that the system changed 12,000,000 feet of air each minute, cooling it with some 25,000 feet of "cold pipes."

As hot weather set in and the system proved its worth, Wolff's work was written up in a slew of technical journals, including *Carpentry and Building, The Metal Worker, Plumber and Steam Fitter, Cold Storage and Ice Trade Journal*, and *The Heating and Ventilating Magazine*. Its health benefits were mentioned in *The Medico-Legal Journal* and the *Journal of the American Medical Association*. It even showed up among the financial intrigues detailed in *The Stock Exchange from Within*, a stockbroker-written book that interrupted its stories of deals both famous and nefarious to insert a lengthy description of the machinery's wonders.

In times to come, the cooling plant of the Stock Exchange would take its place among the marvels of New York, meaning that it would be most noticed when it wasn't working. During a 1911 water shortage the system

The New York Stock Exchange boasted the largest-ever-built "comfort cooling" system, at a time when most people had never encountered such a thing as comfort cooling. (New York Stock Exchange Archives)

was briefly turned off, and this made for a mention in the *Times*: "Passersby in Broad Street yesterday were surprised to see a crowd of brokers standing on the open-air balconies which have hitherto had only an ornamental value. . . . The Stock Exchange has a costly and elaborate cooling and ventilating system, which has been shut down since the shortage in the water supply made it necessary to economize . . . ." Otherwise, it was shown off as a tourist attraction, an occasional interest blurb in papers such as the *Wall Street Journal* ("the most comfortable place in the city these hot days"), and a demonstration of the benefits of Big Money.

More significant, though, were the intangible benefits. Finally, a distinguished (i.e., nontheatrical) and high-profile public building had a system designed not to cool beer or beef or invalids or cadavers but to provide comfort for its inhabitants. And those inhabitants happened to be for the most part members of the privileged class, who hadn't before been able to admit that they might want to be comfortable during the summer heat. Under the socially approved cover of "aiding business productivity," the Stock Exchange system introduced them to summertime comfort—and allowed them to enjoy it.

Wolff immediately began to receive other commissions for indoor refrigeration. But they were few and they came exclusively from the commercial sector, and only from those organizations that could afford the very, very high price tag. Within months after the Stock Exchange opening, Wolff was readying a similar installation for the brand-new home of the patrician Hanover Bank, providing cooling to the main banking room as well as the safe deposit vaults. This design was even more up to date; it kept the cooled air inside the bank, not spilling out onto the street, with the help of a recent invention that was still a great curiosity in New York—a revolving door.

Over the next few years, Wolff designed heating and ventilation plants for a glittering list of clients, including the Cleveland Trust Company, Columbia University's uptown campus buildings, the Plaza Hotel, and the spectacular Fifth Avenue home of the New York Public Library. None of them asked for cooling. The Metropolitan Museum of Art did want it in 1907, but that was openly admitted to be "an attempt to maintain a constant relative humidity, so important for the preservation of paintings." The comfort of museum visitors wasn't the point; after a few summers, it was discovered that the galleries didn't need the system to maintain a healthy humidity level . . . for the paintings. And the cooling was turned off. (The visitors, for their part, weren't consulted.)

Although Wolff had spent almost his entire career in service of the wealthiest clients, he was no elitist; for years he had been a stalwart of the Ethical Culture Society and had taken to the streets to hand out leaflets advertising New York's first free kindergarten. Still, his work was thought of as a benefit to the rich, available only to brokers and bankers.

If he was thinking of any way to bring air cooling to the middle classes, it wouldn't come to pass. In the first week of January 1909, fifty years old, still working and still very much in demand, he arrived home from an evening out and unceremoniously died of a heart attack.

Cooling for Everyone, or at Least Some

This wasn't to say that refrigerating engineers weren't trying to publicize the benefits of indoor cooling in any way they could, even to those who couldn't afford it. The 1904 World's Fair in St. Louis plunged wholeheartedly into displays of the wonders of refrigeration; even the magazine *Alienist and Neurologist* noted this, enthusing, "With the thermometer at

90 in the sun, a snowstorm will spread its amazing relief over the thousands of visitors from every country on earth." Much of that snowstorm was devoted to preserving various agricultural displays, such as North Dakota's life-size butter sculpture of Theodore Roosevelt on horseback or Kansas's scale model (also butter) that depicted a woman operating a cream separator. But Missouri, as host, went in a different direction and tried cooling people rather than dairy products. Its exhibit was the largest state building on the grounds, containing twelve rooms, a 1,000-seat auditorium, and a 76-foot rotunda complete with fountain . . . and it was "cooled by refrigeration." Thousands of visitors ducked in to get a respite from the St. Louis heat, and for many of them it was their first-ever experience with artificial cooling. Still, the building's system received far less attention than it merited. Perhaps comfort-only cooling seemed dull when it was surrounded by attractions that used their refrigeration equipment in much glitzier ways—such as the skating rink that also staged a snowfall each afternoon, or the midway ride "New York to the North Pole," which promised a virtual Arctic voyage by ship, during which "by a peculiar process of refrigeration, the explorers experience all the changes of weather that would be perceptible."

At the same time, there was a surge of interest in the possibility that cold air was a cure for hay fever. The *Transactions of the American Society of Refrigerating Engineers* told the tale of a man, in the midst of a series of allergic attacks, who happened to visit the hold of a refrigerated ship and suddenly found himself not sneezing. Another journal told of a St. Louis man who visited a brewery's beer cooler for "a half hour each day for two weeks" and found his symptoms had disappeared; so many people were interested that the brewery eventually had to repel them with a sign, THIS IS NO SANITARIUM. The refrigeration-men's journal *Cold*, relishing the prospect of additional business, helpfully suggested "the possibility of having in every city a refrigerated club room with telephones and all modern conveniences for the use of hay fever–ites during the term which ordinarily causes them so much agony."

Still, when it came to cooling, there were those who preferred to improvise. In mid-June 1901, Pennsylvania heating engineer John J. Harris was approached by the board members of Scranton High School when they realized that their upcoming graduation ceremonies were scheduled to take place later that week, in the school auditorium, in the middle of a heat wave. With a minuscule budget and only two days in which to come up with a solution, Harris quickly banged together a Rube Goldberg–type arrangement that used ten tons of ice to supply the cooling, the auditorium's fan system to blow air through the ice, large pans of calcium chloride pellets to soak up excess humidity—and plenty of drain piping to slough away the resulting mess. On the night itself the system managed to lower the 90 degree outside temperature to 76 degrees, 1,400 proud parents were made comfortable, and Harris was asked to repeat his installation for the next two years.

Then there was Willis L. Moore, chief of the U.S. Weather Bureau, who in 1902 was in the process of patenting an Apparatus for Cooling, Purifying and Drying Air, a contraption that the *Times* called a "cold stove" and Professor Moore himself called The Nevo—magazine ads coyly told readers, "Spell the name backward." Being a government official, Moore attracted more than the usual amount of publicity; the *Chicago Daily Tribune* photographed him with The Nevo, an eight-foot-tall galvanized object with a jointed pipe protruding from the top (through which hot air would enter) and a small electric fan protruding from the bottom (from which cooled air would emerge). Moore claimed that the device needed no particular attention, "except to be charged once a day with a certain composition." Because he was still waiting for the patent to be granted, he preferred not to identify the composition—which turned out to be nothing more than ice and salt, 250 pounds of it required each day. As this meant that The Nevo had more in common with an ice cream freezer than with actual temperature control, public interest in it, and Moore, quickly faded.

Obviously there was still something of an anything-goes mentality attached to cooling an indoor space. This fact was emphasized by two high-visibility installations. Both were extravagant. Both were designed for New York hotels that were the playthings of their wealthy owners and that opened within months of each other. But these systems were completely different from each other; and one worked while the other didn't.

First up was the indoor cooling plant provided in the Ansonia, a 2,500-room mammoth that opened in April 1904 to become the world's largest apartment hotel. It was also the obsession of its builder, "Copper King" W. E. D. Stokes, eccentric millionaire and amateur inventor. Stokes began the project by envisioning a gigantic version of a French château, bullying architect Paul E.M. Duboy until he had a structure designed precisely to his taste (after which Duboy retreated to France to have a nervous breakdown). Finally satisfied with the design, Stokes then proceeded to bully the builders throughout the construction phase, causing them to

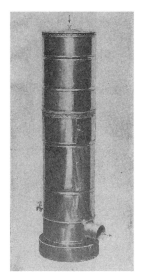
A 1903 ad for The Nevo. Note that it sold for $80 to $350, "depending on size" (in modern terms, from about $1,700 to more than $7,000), and could cost $100 "per summer" (more than $2,000) to operate.
(*Everybody's Magazine*)

walk off the job more than once. Although he was untrained in mechanics, Stokes vetoed the choice of elevators and designed his own improved version; unhappy with the type of terra cotta fireproofing that was available, he went so far as to form a corporation to manufacture exactly the kind he wanted.

At completion, the Ansonia was 800 percent over budget but boasted oddball luxuries that few other buildings had, including a roof-garden farm that provided milk and eggs to tenants, a lobby fountain that was home to a family of live seals, door-to-door message delivery by pneumatic tubes, running filtered ice water in every apartment, complete soundproofing, double-width entrance doors to each suite to allow any tenant a grand piano . . . and for the first time in any hotel, the promise of cooled air in the summertime. *Carpentry and Building* reported, "The air is drawn into the building through screens and filter cloths to the subbasement, where large electric blowers force it over coils, which in winter are heated by steam and in summer cooled by freezing brine, and then discharge it under pressure through the galvanized steel flues which arise through the building." Stokes insisted that this arrangement would keep apartments at a constant 70 degrees. (Some historians described the Ansonia cooling plant much more unconventionally, as a system that forced freezing brine through pipes in the building walls. The pipes were there, installed to provide the chill to refrigerators in each suite. Stokes may have thought that the cold pipes would furnish extra help in cooling the building.)

The bad news came once the building opened; the heating worked, but the cooling system seemed not to function with any efficiency. Perhaps in response to the problem, the Ansonia's rental ads never so much as mentioned it. The ventilation flues would be relegated to heating only, and the brine pipes would be forgotten until 1942, when they were ripped out of the building's walls during a World War II scrap metal drive.

One of the era's more successful cooling installations wasn't all that different from the Ansonia's. But its success was a surprise to a number of people, coming as it did from the imagination of John Jacob Astor IV. They shouldn't have been surprised. Astor wasn't the usual brand of multimillionaire—a Harvard graduate and published science fiction writer, with several patents to his credit, he had commissioned Alfred Wolff to design the heating and ventilation plant for his Hotel Astoria in 1897 (which was exactly next door to his cousin's Hotel Waldorf; the following year the institutions merged, becoming the Waldorf-Astoria). When Astor launched into construction of the Hotel St. Regis four years later, it was obvious that he had paid close attention to Wolff's work but had a few ideas of his own.

The St. Regis opened in September 1904, intended to be the last word in quietly sumptuous lodging, an amplification of the Astoria's opulence and a reflection of Astor's own taste. It didn't disappoint, offering suites with Steinway pianos, a bathroom to every bedroom, silverplated faucets and bath fixtures, magnetically calibrated electric clocks in each guest room, and one of the first central vacuum cleaning systems (an invention so new that it had to be explained in hotel literature: "The maid attaches a small flexible pipe to an outlet found in every room or hall . . . . [W]hen she applies the nozzle of the pipe to furniture or wall, the dust and impurities are sucked down to the basement").

Most surprising of all, where the Astoria merely had been fan-ventilated—and only the public rooms, at that—the St. Regis offered complete temperature control throughout the building. As had been the case with Steele MacKaye and the Madison Square Theatre a quarter-century before, the system's jaw-dropping extravagance was the factor that made it work: $300,000 worth of equipment brought in fresh air to intake ducts located every four stories, filtered it, heated or cooled it, adjusted the humidity, then furnished it to each room in the hotel. Individual thermostats were provided even in bathrooms, allowing guests to choose precisely the temperature they liked.

This was heavily publicized, in a nonchalant way. Because Astor knew well that his target customers were members of the *haute monde*, advertisements that mentioned physical discomforts such as oppressive heat would be off limits. Consequently, the cooling factor wasn't stressed as the main point, and the emphasis was on the fact that the system would allow guests to breathe *pure air*. And because Astor knew that he was *Astor*, it was easy for the St. Regis to broadcast this message in reams of publicity. Full-page ads, disguised as "articles" describing the hotel, each one exactly alike right down to the typeface, ran in papers from Washington to San Francisco, praising the air's purity and featuring the somewhat unsettling line, "Eminent surgeons have sent patients to the hotel for operations, realizing the immense value of its pure air." Even the *London Times* extolled the hotel's air quality. The *New York Times* went furthest of all, devoting an entire page to the ventilation system alone: "When a St. Regis lodger pays $125 a day for apartments in the most expensive hostelry that New York ever had, he or she purchases the privilege of breathing air that has been filtered and prepared . . . being filtered of a barrel of dust a day. . . . [T]he wonderful edifice is supplied with air and heat or coolness the year round." That barrel-of-dust line was repeated seven times in the article (illustrated by drawings that depicted extremely well-dressed hotel guests visiting the Blow Room and the Large Dynamo for Pure Air Apparatus), as well as the facts that the prince of Japan was expected to arrive, was suffering from a cold, the house doctor had prescribed air at exactly 70 degrees, and achieving this would require nothing more than the turn of a knob.

The Hotel St. Regis was a trendsetter, but few builders had Astor's budget. As a compromise, the next years saw a rush of hotels both new and old that touted refrigeration systems of their own, among them Chicago's Auditorium, Congress, Blackstone and La Salle; New York's Ritz-Carlton and Vanderbilt; and Louisville's Seelbach. The problem was that these systems were installed to cool only selected function rooms and

If the *New York Times* was to be believed, high society had decreed that getting a glimpse of the ventilation equipment at the St. Regis was The Thing to Do. (*New York Times*)

restaurants, *not* bedchambers. Advertisements tried to accentuate the positive, shouting lines such as "This is the coolest dining room in the City,"* but the fact remained that if a guest at any of these hotels was going to try to actually get a night's sleep in hot weather, he was probably going to be miserable.

Of course hotels were going to brag about any improvement in summertime comforts, no matter how slight. Office buildings—not high-visibility venues housing big-city power brokers, but workaday business locations—handled it more pragmatically. Before Alfred Wolff had even

*For decades, hotels and restaurants from Boston to Caracas had been making this claim. As it had been based on absolutely nothing, no one really paid attention— until ads began to spell it out as stridently as those for Chicago's Congress Hotel ("Cooled by a New Process of Cold Air"). The Waldorf-Astoria was being comparatively modest when it announced that it had added a fountain to the center of a dining room "through which will run iced water, for the purpose of aiding the hundreds of electric fans in cooling the atmosphere." On the other hand, the Times Square nightspot Rector's used versions of "cool" and "cold" *five* times in a single ad . . . even though Rector's had no cooling equipment.

set to work on the Stock Exchange, in mid-1900 the Armour Packing Company was putting the finishing touches on its four-story Kansas City headquarters. It was a sizeable building for the area, intended to house not only the company's executive offices but also a wholesale meat market on the ground floor as well as additional rental office space. Unhappily, it was going to be located near train tracks (subjecting it to coal dust) and stockyards (subjecting it to walloping odors).

This might have made the Armour Building a less-than-desirable location, but technology had come to the rescue. By early 1901, *Carpentry and Building* reported on

> a system of heating and ventilation heretofore unknown in the West. . . . [T]his is the first building west of the Mississippi that has it.
>
> In the whole building there is not a movable window. The glass is set solid and can neither be raised nor lowered. There are a few transoms that can be moved, but they won't be opened often, for the only purpose they can serve will be to let in the odor which belongs to the packing house district. Fresh air comes in another way.
>
> The temperature and atmosphere of the building are manufactured down stairs in the basement. . . . First [the air] goes through a spraying room where a hundred sprays are throwing water in fine rain. The water is hot or cold according to the season. . . . It literally washes the air and goes a long way toward purifying it besides, changing the atmosphere a few degrees, according to the season.[11]

Other than being described in the *Kansas City Star* and a few builders' journals, the system was ignored by the mainstream press. Perhaps "a few degrees'" change seemed puny.

The Armour Building might arguably have been the first office structure to be constructed with a building-wide air-cooling system. But that distinction was lost, possibly because the building had been designed by William W. Rose, the one-time mayor of Kansas City, Kansas, an architect whose output had been almost entirely local. If there was some status attached to being The First to design an office structure with air cooling, that honor would be seized by the more talented, and more astutely self-promoting, Frank Lloyd Wright.

In 1903 Wright was commissioned by the Larkin Soap Company to design its corporate headquarters in Buffalo. It was his first major chance to strut his stuff, and he responded with a monumental, cathedral-like design, entirely in keeping with the company's overview that employees were part of a close-knit family. Wright then proceeded to make those family members comfortable, apparently without seeking their input, by touches such as natural light provided by a gigantic skylighted atrium and numerous unshaded windows, inspiration provided by company mottoes carved into the stonework, and to provide efficiency he had designed office chairs with three-legged bases, while to make janitorial work easier, other chairs had no legs at all and were bolted to the desks by a swing-away mechanism. Once the building opened in 1906, it turned out that the unshaded windows let in either not enough light or too much (the west-facing windows had to have blinds installed to cut down the after-noon glare, which was said to have displeased Wright); the one-size-fits-all legless clerks' chairs spelled ergonomic disaster for differing body types; and the three-legged seats, which had a habit of flying out from beneath employees as they sat, were promptly dubbed "Suicide Chairs."

More important to employee comfort was the fact that the building was located across the street from the Larkin factory and bordered by train tracks as well, which could make for a smoky, sooty environment; and as one observer pointed out, soot-smudged stationery wouldn't do at all for a soap company. Wright decided to solve that problem by making the building "hermetically sealed" and providing a ventilation plant. In the Larkin employees' magazine, he described the installation:

> By mechanical means the fresh air is taken in at the roof levels, drawn to the basement, washed by passing through a sheet of water sprays (which in summer reduces its temperature two or three degrees), heated (in win-ter), circulated and finally exhausted . . . .[12]

It was a very good idea, even if it didn't quite work. The system—another Acme—proved incapable of lowering the humidity level enough to make the "two or three degrees" reduction of temperature livable. By 1909, additional refrigeration equipment would have to be added to the installa-tion to make the building's atmosphere pleasant.*

The Larkin Building was part of what made Wright's career, but at the time of its construction it got more attention in Europe than in the American press (one notable exception was the *Architectural Record*, whose critic wrote a savage review of the building in which he called it "a monster of awkwardness"). Whether or not it was a failure in terms of comfort for the people who used it, Wright didn't address; such things

---

*Decades later, Wright revisited the Larkin Building with a couple of his appren-tices and "described the air-conditioning system he had devised: air blown over ice to cool it, then circulated through the rest of the building." So much for memory.

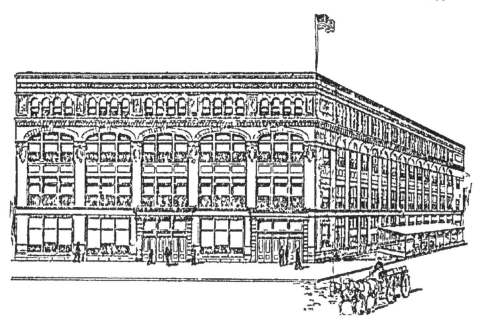

Two contenders for the title of The First Ever Air-Conditioned Building, one nearly unknown, the other world-famous: the 1900 Armour Building (*above*) and the 1906 Larkin Administration Building, with a view of its central atrium (*below*). (*Kansas City Star*/Collection of Jerome Puma/*The National Magazine*)

didn't seem to be terribly important to him. As to its status in the history of air cooling, that came from Wright himself. Throughout his career, whenever he felt it necessary he manufactured his own history. In this particular case, he steadfastly referred to the Larkin Administration Building as "the first 'air conditioned' building in America."

"Let the People Think They Are Cool . . ."

While cooling an office building or a hotel was a welcome prospect, most people still hadn't come to think of it as a necessity. This made sense; throughout their lives, they'd been accustomed to putting up with summertime heat in those buildings. It would require a new and artificial environment to prompt the first public demands for better comfort. And in those days, the newest and most artificial environment that most people experienced was the one that they entered when they descended below the streets to travel by subway.

London had constructed the world's first subway in 1863, a system four miles long that served a total of seven stations. With locomotives that burned coal, and not nearly enough ventilation, as soon as the system opened passengers began complaining of smoke, heat, and soot. So many complaints were made that vents—"blow-holes"—were immediately installed along the line, along with exhaust fans. They were of little help. The Underground became famous for its noxious air—even the fictional Dr. Watson had something to say about it in one of the Sherlock Holmes stories, calling it "abominable." London's reputation would attach itself to subways in general, making for intense and critical scrutiny of each underground transit system that would ever be constructed, anywhere in the world.

(Conditions were quite different but equally unpleasant in the Tower Subway, a 450-yard tunnel under the Thames completed in 1870 whose transportation equipment consisted of a single "omnibus," pulled by cable, capable of holding only twelve passengers per trip, and windowless. The *San Francisco Call* wrote: "The ventilation was bad . . . the air is damp and chill . . . ." Not surprisingly, the omnibus service went out of business after three months. After that, anyone who wanted to use the Tower Subway had to walk through the tunnel. Pedestrians reported that it was creepily dark, the air was still dank, and the walls sweated.)

By the 1890s, underground steam locomotives had been replaced by electric power, which was touted by some as a completely clean energy

KING'S CROSS. INTERIOR

This artist's depiction of the brand-new King's Cross station omitted the
clouds of steam and locomotive smoke that would be trapped by the roof.
(*London Illustrated News*)

source that would solve the problem of ventilation in any transport system. America got to find out in 1897 when Boston opened its own subway, the first in the nation. Its builders had studied the London model in
the hope of learning what not to do, and at its opening there were
installed a number of ventilation towers equipped with exhaust fans. But
two years before, while tunnels were still being excavated, *The New
England Magazine* had reported that much of the ventilation was
expected to take place by another method: "Each car acts as an automatic
ventilator of the tunnel, as it pushes, piston-fashion, a column of air
before it, leaving a vacuum to be supplied by fresh air which draws in
behind."

The Boston subway opened at the beginning of September, and passengers reportedly found the tunnel "refreshingly cool." (The *Kansas City
Journal*, one of the many newspapers that showed up for the event, had

a different take: "Ventilation is imperfectly performed, and the air is cold, damp, and earthy in odor.") The tunnel remained cool, and even cold, through the winter. There was a slight problem in the spring when the temperature rose and warm air condensed on walls and ceilings; passengers riding in the subway's open-sided cars were being dribbled upon. Men with squeegees were sent into the tunnels to wipe down the drip.

Still, compared with London's air-quality nightmares, Boston's annoyances were minor. And when New York decided to build its own subway system only a few years later, the "piston" example was reassuring to engineers and commentators alike. In May 1903, *Traction and Transmission* noted that Boston's exhaust fans were rarely necessary: "We must remark that such artificial ventilation is needed only a few days in the year, the rush of trains in opposite directions being sufficient for the purpose. For this reason no special ventilating arrangements were provided on the New York Subway now in course of construction."

This was true. Other than a handful of "large square openings" to the street scattered along the tunnel, and those only at its northernmost end, there were no provisions for ventilation in the original plans. System designers were confident that station entrances were large enough to draw in all the fresh air that the subway would ever need. At the same time as the *Traction and Transmission* article, the *New-York Tribune* revealed that subway officials had asked for estimates "from various concerns for electric fans with which to ventilate the tunnel." Whatever information was provided, it was apparently ignored.

A year later the system was almost ready for use, and the general consensus was that New York's subway would be just as refreshingly cool as Boston's, no matter how hot the weather. In July, the *Times* reported on an inspection tour of the nearly finished system in an article entitled "Had Coats Buttoned on Tour of Subway":

> [N]ot the least important thing was that citizens who have to be in the city during the heated spell of the good old Summer time will find the subway about the coolest place in Manhattan. Some of those who were in yesterday's party declared that the subway was the final solution of the problem of making New York the most popular of the Summer resorts in America. With the temperature above ground such that men and horses were dropping in the streets, the Mayor of New York and his party rode about under the city with coats buttoned close and in perfect comfort. The temperature could not have ranged much above 70 degrees in any part of the subway.[13]

The *Tribune* was impressed enough to call the subway "a kind of cold storage transit," insisting that "on hot days the cool subway will do a big business." Astoundingly, no one had given a thought to the considerable heat-generating possibilities of the trains themselves—let alone the multitudes of human bodies that would ride in them.

The New York subway opened for business in late October 1904 amid massive civic festivities, nationwide press coverage, and a million people crowding into the system during its first four days. But within hours of its opening, there were complaints of "bad air" and "lack of oxygen," numbers of ladies fainted from the stuffiness, and a prominent physician was quoted declaring, "I should never recommend to any of my patients daily trips in the Subway." From there, various experts spent the next seven months publicly debating the wholesomeness of the underground atmosphere. Health Department officials proved their point by publishing the germ count of subway air, while transit authorities fought back by handing out leaflets printed with the slogan "Subway Air as Pure as Your Own Home." Still, the argument remained fairly controlled.

Until the summer. Starting in early June 1905 the temperature climbed steeply, the first of a string of heat waves that persisted for two months and blanketed nearly half the country. It was the first summer the subway had ever experienced, and riders as well as subway employees were fast discovering that what could be dismissed as "bad air" in cool weather was intolerable during a heat wave. Worse, the nation's press turned its collective attention to New York's subway; while they had spent months being worshipful of the system, now they turned on it angrily. Among dozens of negative writeups, the *Richmond Times-Dispatch* described the plight of a ticket taker at the 116th Street station who was carted off to Bellevue Hospital, driven insane by the heat, while the *Los Angeles Herald* shrieked, "Subway Causes Profound Alarm," attacked "the foulness of the air," and quoted electrical authority Nicola Tesla's worry that the underground atmosphere was so contaminated that the air itself might explode. Closer to home, the *Tribune* found a Boston lady and a London gent, both traveling in the New York subway and both disgusted with its thick heat ("We have pure air in *our* subway!" sputtered the lady as she swept out of the station). The *Herald* sent its reporters into the stations with thermometers and published readings five to twelve degrees hotter than street level; the Brooklyn Bridge station hit 115 degrees. And the *Times* told of a woman who tried to get a stick of chewing gum from a vending machine on the platform, but it was so softened by the heat that it refused either to come out of the slot or let go of her hand, and a thread of gum stubbornly trailed after her, all the way across the platform and onto her train.

Ridership instantly dropped as thousands of passengers opted to use the city's elevated trains, which were gleefully touting themselves as "The Fresh Air Line." And newspapers were publishing letters from readers who had their own ideas for cooling the subway, everything from soaking station platforms with water to equipping each station with "a portable machine" that would somehow cool the air *and* scent it of pine needles. With a public relations nightmare looming, the Rapid Transit Commission called a meeting; as one member grumbled (when he thought no one was listening), the point of the meeting was "to stop the 'kicking' of the public." A group of experts, Alfred R. Wolff among them, had already recommended the Boston model of ventilation shafts and exhaust fans. This was ignored. Instead, it was suggested that a number of smaller electric fans be installed. "They may not do the business, but the people will see the fans going and that will be something," said one commissioner. When someone questioned the usefulness of this idea, the head of the commission snapped (when *he* thought no one was listening), "Let the people think they are cool and they will be cool."

The humor magazine *Puck* satirized a man's attempts to cool off during a heat wave, which include an extremely ill-advised subway ride. (Library of Congress, Prints and Photographs Division)

But there was also interest in "some system of clearing and, perhaps"—and here was a word spoken aloud probably for the first time in a meeting that discussed a public project—"of *cooling* the air . . . ."

Not the cars: curiously, the stations themselves. A week after the meeting, a number of newspapers and even *The Street Railway Journal* reported, "Refrigerating plants and forced-air apparatus are to be installed in the subway. . . . Tests are to be made as soon as possible at two or three of the stations, and then, if the apparatus proves efficient, the entire subway will be equipped with the plants."

The local press didn't buy it. The *Times* editorialized, "[T]he Rapid Transit Commission propose[s] to install an ammonia refrigerating plant in the Subway to cool it. . . . As a matter of fact they have no such intention." And the *New York American* jeered, "Great improvements in subway ventilation are promised for next year. Hold your breath until 1906."

Sure enough, the public had to do exactly that. For the rest of 1905, the Brooklyn Bridge and Grand Central stations made do with three dozen small electric fans. Much of 1906 was spent in purchasing large-scale exhaust fans, hacking out new ventilation shafts, and installing gratings in sidewalks (which most people initially despised; from nervousness or modesty, they hated even the thought of walking over them, especially when passing trains kicked up gusts of air). All the while, there were various reports of plans to install refrigerating equipment at the Brooklyn Bridge, 14th Street, Grand Central, and Times Square stations.

But there were snags. That summer, the *Municipal Journal and Engineer* snorted, "The fans for cooling the subway have been lost somewhere in a freight car between this city and some place out West. It is hoped, however, to have them found and installed, ready for the winter." And while the refrigerating equipment was repeatedly promised, and in print, it showed up at the Brooklyn Bridge station only toward the end of August. The system was an unusual hybrid for its time, using giant blowers to force air over coils that circulated cool water drawn from two artesian wells. While it was lauded in print, it wasn't very much help; even the dryly official record *The Air and Ventilation of Subways* admitted, "Although the regular passengers at this station took up positions immediately under the openings when the plant was operated, the effect seems not to have been perceptible to the senses at other points in the station."

One report claimed that $350,000 had been spent to improve subway air, $45,000 of it on the Brooklyn Bridge contraption alone. After that,

The New York subway system's contribution to passenger comfort:
5,000 General Electric ceiling fans. GE was obviously proud of this job; they
ran their own advertising in trade journals, pointing out that it was the
"largest installation in the world . . . These thousands of fans are giving
thoroughly satisfactory service under the most extraordinary conditions."
(*2005.61.129 IRT MUDC Hedley Car, an HV motor*—Vincent Lee Collection,
Courtesy of New York Transit Museum)

little else would happen in New York to provide summertime comfort for underground passengers. The attention shifted from stations to subway cars themselves, which had been receiving most of their air through open windows or meager ventilation openings. In 1910, a few were provided with "experimental" ceiling fans, a smaller version of the ones found in any saloon. Even though the *Tribune* sighed, "Electric fans will not improve the quality of the air," and even though their open-bladed construction made them a bad idea in the subway car's close quarters ("Tall people had to be extremely careful," remembered one rider), there was no alternative. They would be installed in nearly every car on the line, becoming the default cooling system in the New York subway for the next seven decades.

As to the "refrigerating" setup at Brooklyn Bridge, it chugged along with its reported minimal effect until 1916, when it overheated on a hot day and caught fire. Repaired, it continued for another decade or so until new construction in the area caused the artesian wells to run dry. At that point it was stopped without fanfare.

3

For Paper, Not People

In the first years of the twentieth century, mechanical air cooling had reached a very strange crossroads in its development. In theory, it was seen as the most modern example of scientific progress, a flashy machine that could conquer the weather. But it was almost too modern for its surroundings. For instance, at the same time that venturesome heating-and-ventilation men might be attempting "refrigerated rooms," they were being cautioned to take into account the extra heat in those rooms that would be generated by . . . gaslight.

There was another, more serious problem; it rarely worked. Yes, chilling milk in a warehouse with machine-made cold had become commonplace. But trying to cool a room had proven to be impractical, ridiculously expensive, and error-prone, a jigsaw puzzle that hadn't been assembled correctly. Some well-read engineer had jovially picked up the old English term *coolth* to describe what came out of the machine if everything was working well. But far too often, everything didn't work well and *coolth* didn't happen. There were few "refrigerated rooms" in existence—and the majority of them had proven to be maddeningly temperamental installations, often too warm, or too humid, or both at the same time. In order to be seriously accepted as a useful technology, this branch of the refrigeration field would need a genius to solve its last remaining glitches.

And that genius would need to be a front man. Members of the public in the early 1900s loved to view themselves as "go-ahead" twentieth-century types, but their attitude was still unbendingly Victorian, and they were rattled by the era's flood of newfangled scientific contraptions. The devices that were most successful seemed to be the ones that were linked to a reassuring human face. As Bell had fathered the telephone, and Edison had become the representative of electric light, mechanical cooling needed a persona: someone who would promote it with unflagging enthusiasm.

It would find the genius, as well as the front man, in Willis Haviland Carrier.

Born a farmer's son in 1876 and fascinated by mechanics of all kinds, Carrier worked his way through Cornell University as a scholarship student, taking any odd job available in order to eat. Awarded the degree of Mechanical Engineer in Electrical Engineering in 1901, a month later he was on his way to a position with the Buffalo Forge Company, manufacturer of fans and heaters—and fatefully, industrial drying equipment—at the bare-bones salary of ten dollars per week.

Carrier proved to be an extremely valuable employee, and a persistent one, too. Buffalo Forge had begun life in 1878 by manufacturing blacksmithing equipment and had fallen somewhat haphazardly into the ventilation game when it found its blacksmiths' blowers being remodeled to bring air into factory spaces. But the leap into the twentieth century was proving to be a difficult one. The company's installations were often too-much-or-not-enough, designed with the help of the same faulty calculation tables that had plagued Alfred R. Wolff years before; this was resulting in extra work and client annoyance. However, the new breed of engineer sneered at this kind of guesstimating as the "cut and *try* method," and Carrier refused categorically to operate by any such rules. He insisted that Science had to be applied to the problem, challenging every step of the design and installation process, even testing the air friction inside duct pipes. Members of the old guard at Buffalo Forge were exasperated—until they realized that Carrier's tinkering had saved the company roughly $40,000 in a single year by heading off costly installation blunders. This made it easy for him to talk his way into setting up Buffalo Forge's Department of Experimental Engineering in 1902. (It was easy for the company as well. They gave him no title and no laboratory space.)

As it happened, Carrier's timing was perfect. Barely a month later, he would be given the assignment that would change his life.

Brooklyn's Sackett-Wilhelms Lithographic and Publishing Company had become known for high-quality color printing, especially of the national humor magazine *Judge*. But they were beset by quality issues. Color printing required a single sheet of paper to make several passes through a printing press, each time printing with a different color of ink; high humidity caused paper to swell; and even the slightest change meant that successive passes through the press would be unable to "register," producing not a sharp color image but a blur. Because the summers of 1900 and 1901 had been brutally hot and humid, mountains of paper stock had been ruined—and there had been days when no printing could be done at all, a serious handicap when deadlines were involved. Now, the summer of 1902 was heating up as well. Scared into action, Sackett-Wilhelms contacted Buffalo Forge for their industrial drying expertise. And Buffalo Forge executives took the problem to their newest Experimental Engineer to see if anything could be done about the humidity problem.

Carrier would later remember that, at the time, "I had never heard of any such thing as air humidity."

Still, he set to work, starting with nothing but a floor plan of the Sackett-Wilhelms building, a slide rule, and a handful of temperature–humidity charts supplied by the U.S. Weather Bureau. Someone recommended calcium chloride to absorb humidity, but there would have to be a way to use it. He wrote, "I rigged up a burlap cloth on two rollers. A fan pulled air through the cloth. The cloth was wetted continuously with a saturated solution of calcium chloride brine. Everything except the fan was manually operated . . . a man to dip brine from a barrel and pour it over the cloth, and a man to turn the rollers." After days of blowing air, dipping brine, and charting moisture levels, Carrier decided that the system was not only completely impractical but also commercially useless. Monstrous amounts of calcium chloride would be needed; it would be a very high-maintenance operation; and, worst of all, the air blowing through the burlap deposited salt crystals on nearby machinery, which would corrode the metal. As he recalled, this knowledge resulted "only in ruining two perfectly good pairs of expensive shoes worn by my two assistants just out of college."

Carrier promptly abandoned the idea of a chemical fix and hit pay dirt when he fashioned some existing Buffalo Forge equipment into a system that resembled a tiny version of Wolff's Stock Exchange setup, using an industrial fan to blow air over some borrowed steam coils that were filled with cold water. With the air at just the right speed and the water at just the right temperature, the excess humidity obediently condensed on the coils and the air came out cooler . . . and, theoretically, as dry as any printer would need it to be.

Translating the theory into a set of blueprints, he presented the drawings to Sackett-Wilhelms on July 17, 1902. They approved the installation. (The date as well as the job would later become famous, but at the time it was completely ignored except for an item in the August 23 issue of *Electrical World and Engineer*. The article reported that Sackett-Wilhelms was having Buffalo Forge install generators and "three engines, which are to be used for lighting and power purposes.")

The system was up and running by early autumn. Buffalo Forge was happy with the "excellent results." At first Sackett-Wilhelms was happy, too, but within a year or so the system had been deemed inadequate, modified with additional equipment, and ultimately removed.* And Carrier himself "realized that the design was not the final answer for controlling the moisture content of the air."

He was right. Various methods had been developed to cool air; but a way to control humidity—accurately and *automatically*—hadn't yet been invented, not to anyone's total satisfaction. Even that installation of installations, the system at the New York Stock Exchange, had to rely on human help. The *New York Times* noted that visitors "are sometimes surprised to see a man weaving his way through the crowd with head thrown back, sniffing the air, while with an extended hand he appears to seize

VENTILATION

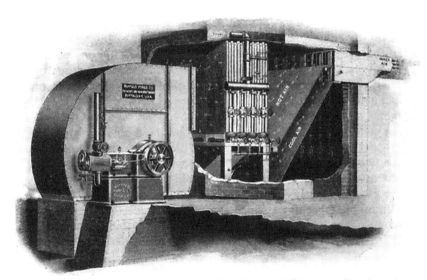

Three-Quarter Housing Fan, Left-Hand Top Horizontal Discharge, Blowing Air Through and Underneath Heater into Brick Plenum Chamber

At a casual glance, one roomful of ventilating machinery looked much like another—but Carrier's installation was about to start a change in the way that people would view the threat of summer heat.
(Courtesy: Carrier)

*It wasn't removed completely. The Brooklyn building survives to this day, and so does some of the brick ductwork that conveyed the dried air up to the main level.

large pinches of the atmosphere. . . . [H]e is the expert in charge of the ventilating plant. . . . His senses have become so completely adjusted to the work that he can estimate the humidity of the air by feeling. . . ." That arrangement might have been all right for the Stock Exchange, but Carrier knew that if every system had to have a professional air-sniffer holed up in the back room, no one would want the system.

He began to work toward a way to solve the problem, and the answer came to him within weeks, as he was waiting for a train in Pittsburgh on a foggy night. Looking at the mist in the air and the condensation on the train windows, he realized the solution in a flash: *Fog itself* could be used to gather moisture and force it to condense (basically, what happens when fog turns to rain). An artificial fog could be created by adapting the idea of the air washer. By controlling the temperature of the air that blew into it, and the temperature of the water that created the fog—a warm fog would create more humid air, a cold fog drier air—he could produce any level of humidity at any temperature. On paper, the idea worked perfectly.

Starting with an off-the-rack air washer, a collection of miscellaneous Buffalo Forge parts, and an array of fine-mist nozzles that had originally been designed to spray insecticide, he constructed a prototype machine. By early 1904 he was patenting it as an Apparatus for Treating Air. Outside of Carrier's immediate circle, the idea drew sneers rather than interest: One biographer wrote that the mere idea of using *water* to lower humidity was "greeted with incredulity and, in some cases, with ridicule." Nevertheless, a year later Buffalo Forge was offering the system as the Buffalo Air Washer and Humidifier.

Once again, the timing was perfect. A North Carolina textile engineer named Stuart Cramer had been working on devices that could increase the humidity in textile mills—overly dry air caused threads to break, thus slowing production. In past eras, mill operators had resorted to sprinkling water on factory floors, a dangerous idea for workers walking around high-speed machinery. Later, there had been some atomizer-like devices that sprayed water into the air, but they spattered droplets on machines and caused rust. Cramer developed a system to add humidity to factory air in a finer, more controlled way. And he gave it a name: "air conditioning."

Carrier—who felt that air could be "conditioned" just as much by removing moisture as by adding it—thought the phrase had a certain ring.

As it turned out, the textile industry wasn't the only one that was finicky when it came to factory air. Nearly every other manufacturer knew that certain weather conditions were good for production, and other conditions disastrous. Soap, leather, glue, macaroni, and pharmaceutical compounds needed to be dried, but each at its own particular rate; bakeries and tobacco producers needed moist air, but not too moist; chewing gum and chocolate makers were so threatened by heat that many of them preferred to shut down entirely during the summer. When it came to production schedules' colliding with the wrong type of weather, waiting it out was an expensive proposition. For the first time, Carrier's equipment was offering a real alternative.

Carrier himself wrote the copy for the Buffalo Air Washer's first catalogue. The magazine *Compressed Air* paraphrased it: "This apparatus removes all dust and smoke from the air entering a ventilation system; regulates the humidity; reduces the temperature in summer . . . requires no attention or adjusting." The first year the system was on the market, only one unit was sold. But after that, word spread quickly and, better still, in print. The *Louisiana Planter and Sugar Manufacturer* wrote during the summer of 1906, "The hot summer days now prevailing make one wonder why ventilation with cool currents has not been thoroughly well exploited before these days. The Buffalo Forge Co., by this air washer and humidifier, is . . . solving the problem of ventilation. . . ."

Gradually, the Buffalo Air Washer took off. So, for that matter, did Willis Carrier. He soon was made the head of the company's new Engineering Department; and this time, he got a building of his own. By 1907, Buffalo Forge acknowledged this success by creating a subsidiary, the Carrier Air Conditioning Company of America. It was quite an achievement for a man who had earned a reputation for complete absentmindedness, so engrossed in solving the puzzle of air conditioning that he became known for leaving his wife on the station platform while he boarded the train (and more than once), ordering a multicourse lunch but being too absorbed in calculations to eat a bite of it, and heading out on a trip with a large suitcase in which he had packed nothing but a handkerchief.

If there was a problem, it was that this sensation was going unnoticed by people outside the industrial-ventilation world. In the Buffalo Air Washer's initial catalogue, Carrier had made a point of recommending the system for "cooling theaters, churches, restaurants," but no one went for it. In fact, the very first Buffalo Air Washer sold had gone to the La Crosse (Wisconsin) National Bank . . . but specifically to *clean* the air, not to cool it. It wasn't until 1908, when the Hotel Astor installed a system to clean the air in a dining room and discovered on a hot day that the temperature happened to be ten degrees lower than anywhere else in the building, that the hotel installed another unit in its basement-level Indian Grill Room and became the first customer anywhere to take advantage of the Buffalo Air Washer's cooling capability.

Obviously it was going to take a while for Carrier's product to reach the general population. Still, it had been an astounding decade for Willis Carrier. In 1911, he went public with his research when he presented it under the title *Rational Psychrometric Formulae* at a meeting of the American Society of Mechanical Engineers. The paper provided a whole series of calculations that could allow any ventilation designer to tailor precisely the equipment to the job: utterly unglamorous information to the casual observer, but a Rosetta Stone for engineers.

In the world of air conditioning, the *Rational Psychrometric Formulae* instantly made Carrier into the front runner. In truth, this wasn't much of an achievement, as Carrier had the field nearly to himself. Few companies made air washers, and fewer were promoting them as anything but nuts-and-bolts factory equipment. The Webster, the Bickalky, the Sturtevant Air Washers advertised their wares with little excitement: One manufacturer claimed only that its product had "been widely adopted in laundries for this purpose, and it has been found of great value in furniture factories. . . ." If more drama was wanted by the purchaser, there was the old-fashioned hokum of a magazine ad that appeared in trade journals at the same time:

AVOID DEATH
BY
IMPURE, FOUL AIR
THOMAS' ACME AIR WASHER

Washington's Hot Air (Part IV)

The summer of 1911 was pure heat-related hell. More than 2,000 people died in the United States alone; even Europe was affected, with temperature records breaking all over the continent. And the nation's capital reacted with continued attempts to keep rooms cool . . . by using ice.

Alexander Graham Bell, a Washington resident, was one of them. As he told reporters, a recent worldwide tour had impressed him with the fact that people who lived in extremely hot climates seemed to do very little about cooling themselves. Returning to America, he was determined to cool at least part of his own home.

His method began with a large insulated ice chest in his kitchen. Into the chest ran a pipe, with a small fan bringing in air. From the chest extended another pipe, wrapped in asbestos, which ran through the kitchen wall, turned downward, and extended into the basement, ending in the household's tile-lined "swimming tank" (drained for the occasion). A few pieces of furniture, a couple of rugs, an electric lamp—along with a stepladder to climb down easily—and as soon as Professor Bell flipped the switch on the fan he had a comfortable reading room, with a supply of cool air flowing into it. *Popular Mechanics* admired the setup: "The thermometer never went above 61 degrees during the summer."

If it seemed that Bell had borrowed the idea from Willis L. Moore and his Nevo, he probably had. Moore was a fellow Washingtonian, and both men saw a good deal of each other as officers of the National

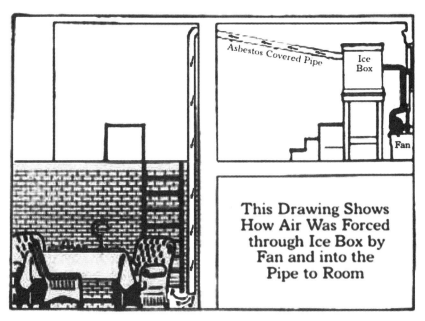

A diagram of Bell's "Ice Stove," which was written up in every publication from the *Bell Telephone News* to *The Standard Illustrated Book of Facts*. The *New York Times Sunday Magazine* was astonished: "There are numberless cooling devices, full of complicated coils and chemicals with hyphenated names that require a professional chemist to operate. But anyone who can dump a cake of ice into a box and turn a switch can operate this cooling stove of Prof. Bell's." (*Popular Mechanics*)

Geographic Society. Moore would return the compliment in 1924, when he told the *Washington Post* that people should live in basements during the summer, "just as Sahara Desert tribes burrow into earth to escape the heat."

While Bell's idea made for very interesting reading, most homes weren't equipped with swimming tanks of their own. One of those unfortunate residences was the White House of President William Howard Taft. If anyone needed a way to cool his surroundings, Taft did: He weighed more than 300 pounds and was perpetually exhausted from sleep apnea, which became worse in hot weather. When he had moved to Manila in 1901 for a three-year stint as Governor-General of the Philippines (he was quoted as saying that there were only three places hotter than the Philippines: his hometown of Cincinnati, Washington, "and the nether regions"), one of his first priorities was to install ceiling fans in his residence. When his wife came to join him, she was appalled at the very sight of them. To her mind, ceiling fans belonged in an ice cream parlor.

Manila was hot, and so was Washington; by the time the Tafts had become the First Couple of the United States, the President was made miserable by the heat and decided to have some sort of cooling apparatus installed. But there was a political side to such a decision; any equipment purchased by the President for his own summertime comfort—possibly paid for by taxpayer dollars!—could result in partisan finger-pointing, along with accusations of hedonism, from constituents and press alike. So the best that could be mustered for the nation's Chief Executive was a makeshift system that ignored Alfred Wolff, Willis Carrier, and every bit of current technical knowledge. The White House attic was rigged up in cooling-machine style, circa 1880: racks of ice, fans blowing over them, the resulting cool air routed through the heating ducts. Apparently it was useless. In fact, when Taft was confronted with the necessity of hosting a state dinner that July, he opted not to hold it in the State Dining Room but *on the roof* of the West Wing.

Perhaps the White House was a structure that didn't take kindly to major changes. However, the West Wing itself had been built only a few years before and could be altered without fuss. And as it happened, that building underwent an extensive reconstruction in late 1909, giving it not only the first Oval Office but also a "cold-air plant" operating in a basement vault beneath it:

Apparatus is provided to force the cold blasts from the vault into the President's room over chloride of lime, which removes all dampness without raising the temperature. It is asserted that the equipment will control the temperature in the President's room during the hottest days of summer to a degree that the Chief Executive will find comfortable.[1]

This cold-air plant was another technological throwback, devouring a ton of ice each day to provide the cold. But because it was constructed on a vastly smaller scale, servicing only the Oval Office—as well as a soundproofed, custom-built phone booth for long-distance presidential calls ("Not only will the booth accommodate satisfactorily so big a man as President Taft but there is ample space for a stenographer to stand beside the President's chair inside the booth")—it appeared to work. When it was started up in the summer of 1910 the President seemed pleased; in fact, when members of his Cabinet were called for a midsummer meeting, he issued "an executive mandate" to move the session from the 90-plus–degrees Cabinet Room to the Oval Office. But *Forbes* magazine reported that Taft was ultimately disappointed with the results. At that point he gave up and "resigned himself to the inevitable."

Things were no better at the Capitol, whose legislators had abandoned the idea of demolishing walls to get windows and outside air but were still trying to find a way to cool the building. Some observers found it entertaining: The *New-York Tribune* reported in mid-1910, "Several years ago the Senate chamber was equipped with an apparatus to refrigerate the air." Even though that "apparatus" had never been installed, factual details didn't bother the *Tribune*, which continued, with rib-poking humor, "It was never used, however, because some physician reported that such artificially cooled air would be fatal to men of sixty or more."

Fatal or not, lawmakers bravely decided to risk it, and only a month later an appropriation of $72,000 was voted to install cooling equipment for the houses of Congress. But, given the age of the building and the extremely odd configuration of the rooms, the job wouldn't be easy. By early 1911, the *Congressional Record* reported that a sum total of 25 applications had been received from ventilation firms, only six of them had sent engineers to look over the rooms . . . and only three of those had submitted actual estimates. One of those firms was Thomas & Smith—which, perhaps cautiously, offered the Acme Air Washer and Cooler "to be erected in connection with any refrigerating apparatus installed."

In the end, no action was taken. And the House Committee on Ventilation and Acoustics was disbanded.

That June, the *New-York Tribune* ran a full-page tongue-in-cheek article, "How Congress Tries to Dodge Great Heat of Summer in Washington." It listed lawmakers' methods for summer survival: lemonade stirred up by Senate cloakroom attendants, perhaps a refreshing bath ("bathing rooms" had been in the basement since the 1870s, and it wasn't uncommon for congressional members to disappear in the middle of the day for a quick restorative plunge). Most effective of all was a ride in the twelve-seater jitney provided for travel in the Capitol Subway; the tunnel was far enough underground that it was perpetually cool. As to any talk of refrigeration equipment, the article provided yet another red herring when it mentioned that "an ultra-modern plant" for cooling would be installed "in another year."

When Woodrow Wilson succeeded Taft in 1913, he discovered his predecessor's cooling system. This precipitated a small scandal that summer, noted the *Report of the Proceedings of the Ohio State Bar Association*, citing an article

> which was sent out through a Washington News Bureau, unjustly criticizing President Wilson because ice was being used to reduce the temperature of his business office in the White House, and stating in flaring [sic] headlines that $65.00 of the good people's money was being paid out every day for ice to cool off the White House. The article then went on to describe President Wilson as enjoying himself in his cool office while members of Congress and other officials, to say nothing of common citizens, who paid the taxes, had to take pot-luck with the thermometer standing above 100 degrees. . . . President Wilson was so hurt by the item that he immediately had the cooling system torn out.[2]

Manufacturing . . . Weather?

Ignoring Washington's ice-cooled debacle, the Carrier Air Conditioning Company of America was steadily expanding its customer base. In 1913, the company even made its first residential installation.

Charles G. Gates, son of barbed wire tycoon John W. Gates, was a twenty-something rake who had just inherited his father's $15,000,000 fortune. The family was notorious for headline-grabbing flash; as the elder Gates's nickname had been "Bet-a-Million," Charles became known everywhere as "*Spend*-a-Million." And he did exactly that when he married a Minneapolis debutante and offered to build her a "cottage" in her hometown. The 38,000-square-foot *palazzo* took up three city lots and was done up in true robber-baron style, with a $50,000 pipe organ, gold doorknobs, gold plumbing fixtures, a gold-covered ceiling, at least one gold-plated bathtub, a ballroom lit by 802 lightbulbs . . . and Carrier air conditioning. The machine was the same unit that a "small factory" would use and by any standards a basement-filling behemoth: seven feet tall, six feet wide, and twenty feet long, measurements that didn't include the separate refrigerating unit.

As fate would have it, Gates wouldn't get to enjoy his new surroundings. With the house not yet finished, in the midst of a hunting trip Gates died—of heart failure or apoplexy or appendicitis, depending on which source was quoted. His widow lived in the house for less than two years before she remarried and headed to Connecticut. Another owner visited the house from time to time but never lived in it. It was demolished in 1933, and the best guess is that the air conditioning had never been so much as turned on.

Not only the largest house ever built in Minneapolis but the first air-conditioned house anywhere, the Charles G. Gates house stood for only nineteen years before it was demolished. (Photo: C. J. Hibbard, Courtesy of the Minnesota Historical Society)

Even so, the Gates mansion would become famous as The Very First Air-Conditioned House.* While the title might be debated in light of those 1890-era "refrigerated living rooms," it said plenty about Willis Haviland Carrier's increasing recognition factor. Also his success. Sales of Carrier equipment were steadily climbing: 63 contracts in 1912, 93 in 1913, 130 in 1914 . . . .

. . . at which point the Buffalo Forge Company decided to close the Carrier Air Conditioning Company of America.

It was a case of twisted logic. Buffalo Forge had been supplying equipment to ventilation engineers and began to worry that the Carrier operations might be viewed by those engineers as direct competition. On top of it, Europe was heading into war, and while the United States wasn't thinking of involvement, money men throughout the country were sternly advocating a program of belt-tightening. Top executives at Buffalo Forge agreed. Carrier's job would be safe, he was told, but nearly all of his air conditioning employees would be fired.

Instead, Carrier resigned from the Buffalo Forge Company. With six other engineers and a total start-up capital of $32,600, he formed the Carrier Engineering Corporation in 1915. At first, it was a very economical organization; even the office furniture was secondhand. A secretary remembered, "We had two wicker chairs for visitors, and Mr. Carrier's friends would ask him if he had swiped them from a tavern."

Nevertheless, in the firm's first year it won forty contracts. Manufacturers realized that a Carrier installation was far more thorough than most of its competitors', as Willis Carrier planned systems by taking into account not only the size of a building but also its location, the number of bodies that occupied it and the number of hours they would be there, the power demands of the machinery, and even the building's construction. "A wall composed of two-inch pine boards covered with building paper and corrugated iron, with a difference of 50° in temperature between the two sides," might have seemed trivial and obsessive to another engineer. But Carrier believed that air conditioning was not only

---

*But Gates's thunder may have been stolen a few years before when typewriter inventor and eccentric millionaire James B. Hammond decided to spend his last years at sea. To do so, in 1911 he commissioned a yacht, *Lounger II*, which was equipped with "a refrigerating plant to keep the air in the cabins and staterooms at a cool, dry, even temperature, climate or locality notwithstanding." Only a bit luckier than Gates, Hammond got to enjoy comfort-cooled yachting for two years before he died in the middle of a voyage.

a science but an *art*, and his fascination communicated itself to prospective clients. Clients, in return, remembered not only the company but its president. If the field was gaining a human face, it was Willis Carrier's.

With youthful exuberance, the new company did something that hadn't been done before in the nuts-and-bolts world of engineering: It vigorously marketed air conditioning with a sense of fun. A prospective customer would receive not the usual price-list-and-parts catalogue but a booklet specifically tailored to his own business, loaded with photos of other Carrier installations and testimonials from satisfied customers. The tailoring was an extremely smart idea: Seeing a list of testimonials and installations that came from a customer's own competition was quite likely a strong inducement to invest in a Carrier system.

Before long those booklets morphed into *The Weather Vein*, a free "monthly magazine" featuring a beaming cartoon character named The Mechanical Weather Man, "Mech" for short, whose control-valve body was emblazoned with the phrase "EVERY DAY A GOOD DAY." In cartoon-strip format and in rhyme, Mech visited businesses ranging from spaghetti factories to drug companies, effortlessly solving their air-quality problems:

> The Messrs. Pill and Powder, convinced by "Mech's" display
> Of wisdom, put it up to him to make their business pay
> "Mech" waves his magic wand, then, conditioning the air,
> And all their costly chemicals which used to cause despair
> Are standardized exactly and made superior,
> While everything competitive becomes inferior.

Mech was obviously a load of laughs; still, the fun was aimed at a business audience. While Carrier brochures advertised "Humidifying, Dehumidifying, Cooling, Drying, Air Washing, Automatic Temperature and Humidity Regulation," those benefits were enjoyed almost exclusively by the commercial world.

The military world took advantage of those benefits, too, when the United States entered the Great War in 1917 and air conditioning proved itself indispensable to munitions production. Manufacturers had always feared static, which could ignite flash powder caught on a worker's perspiring fingers and trigger an explosion, fuse assembly was at the mercy of humidity, and the explosive compound ammonium nitrate had to be produced in stable surroundings. Some gigantic installations—one of them coming in at $6,000,000, "the largest air conditioning plant, we

Mech. (Courtesy: Carrier)

believe, that was ever constructed," according to Carrier's partner J. I. Lyle—solved the problem. By war's end, manufacturers seemed finally to agree that air conditioning was an expensive investment, but one that paid for itself quickly. And the term "process air conditioning" sprang up to define systems used for business and factory production.

However, "comfort air conditioning," the other side of the coin, was still nearly unknown. Cooling machines and electric fans were all well and good, but when the thermometer climbed, most people were still operating under the Victorian tough-it-out creed. *American Medicine* had tried to point out that hot weather was not only uncomfortable but also deadly, using the bluntest possible language: "All the sick babies slaughtered by a heat wave could be saved by putting them in a cool room but unfortunately we use the cooling machinery only to keep their poor little bodies from decay after the heat has killed them. The dead room should be the living room." But as *American Medicine* was a professional journal, preaching to the converted, it didn't create any particular sensation.

Almost in response, the 1919 Carrier opus *The Story of Manufactured Weather* showed an installation at the Baby Incubator of the Allegheny County Hospital. As well, there were photos of other such installations at the Fifth Avenue department store Lord & Taylor and an unnamed movie house . . . and a suburban two-story white Colonial Revival house, beautifully tended. The caption read:

> There's no place like Home, the poets say, but do you have to go away from yours during the Summer, because the Weather is unpleasant? Then your Home is only half a Home, isn't it? For the same reason which led you to install heating equipment to make the Winter comfortable, you should also install cooling equipment to make the Summer, the Delight-Time of the year, equally as comfortable. Manufactured Weather, in your Home, would make "Every day a GOOD day." Winter and Summer. How about it?

But only two pages before, the text acknowledged, almost in defeat, "The non-industrial applications,—residences, schools, churches, theatres, office buildings, etc.,—have been comparatively few,—because the Average Man is a peculiar animal in the way he dissociates the principles he employs in his business and those he uses in the conduct of his home, his church, or even his amusements."

Obviously air conditioning wasn't going to endear itself to the public via the factory. Another route would have to be found.

4 *Coolth*: Everybody's Doing It

It had been fifteen years or so since the term "air conditioning" had been born. And, enthusiastically or reluctantly, a whole range of businesses had accepted the fact that it was essential equipment for smooth year-round operations. But to the average person, air conditioning was a strictly-business proposition. Logical enough: Unless they worked in air-conditioned factories, or happened to visit the World's Fair or the New York Stock Exchange, or entered one of the handful of air-cooled hotels or banking houses, it was unlikely that most citizens would have encountered comfort cooling at all, much less looked for it in their daily lives.

That was about to change, if slowly.

For that matter, the very mindset of the nation was preparing to change. The impact of the Great War had gone a long way toward eradicating the last vestiges of Victorianism, and the stiff-upper-lip attitude of yesteryear—especially the idea of self-denial as a form of character building—was fading. Instead, the phrase "personal comfort" seemed to be popping up in the media everywhere, linked with bathroom fixtures, underclothing, church pews, Fuller brushes, street-sweeping schedules, changing-room lockers, living room seating, even house lots in Meriden, Connecticut. And modern technology was promising to help: Immediately after war's end, General Electric advertised, "At the touch of a button innumerable services are performed for man's personal comfort and convenience." Personal comfort, at the touch of a button—a new concept, and an irresistible one.

Mechanical cooling, *comfort* cooling, would become a part of the wondrous new existence that was available through technology. The coming decade would be known as the Roaring Twenties, a time of upheaval that would see more changes in people's daily lives than at any other time in history. As one of those changes, members of the public would be exposed to air conditioning en masse; more significantly, they would not only get used to it but finally would come to *demand* it in their lives, as they never had before.

Hooray for Hollywood

Once again, the idea of comfort cooling would be slammed home to the average citizen via the theater.

Not the "legitimate" theater, however. Since the cooling-machine furor of the 1890s, the appeal seemed to have worn off. By the turn of the century, few theaters were bothering to install further equipment, and those that did were still using ice. There were exceptions here and there: The Teatro Municipal of Rio de Janeiro opened in 1909 with an auditorium cooled by a system that forced air over 50,000 gallons of chilled brine. And when the New York Hippodrome opened in 1905, the 5,300-seat monster, touted as "the world's largest theatre," had spent $50,000 on a refrigeration plant. But once again, the equipment was mismatched to the building; in the summer of 1910 *Ice and Refrigeration* pithily reported, "The mechanism which was installed for cooling the New York Hippodrome, at Sixth Avenue, New York City, is to be removed and the space converted into dressing rooms. The cooling apparatus, it is stated, proved to be inadequate."*

Only months later, the vaudeville producer Jesse Lasky made his Broadway bow when he opened the revue *Hell* in early 1911 at his brand-new theater, the Folies Bergère. Dubbed "More Parisian than Paris" and cultivating an ultra-naughty reputation, the Folies was contrived to be one of the most expensive nightspots in New York. It scored a number of firsts: the first dinner theater, the first regular midnight performances, the first cabaret show (such a new idea, in fact, that ads had to explain it was "pronounced *cabaray*"). And as part of its million-dollar price tag, Lasky decided to ensure his patrons' comfort by dedicating $40,000 of it to cooling equipment—purportedly the first air conditioning ever in a Broadway theater. His choice: an Acme Air Purifying and Cooling System.

While the *New-York Tribune* asserted that the Folies Bergère was "comfortable even on the hottest days," apparently they fibbed; the engineer who had been responsible for the system later acknowledged, using a very odd choice of words, that "only the inefficiency of the apparatus saved the installation from being unbearable." And whether it was because of the unbearableness or the entertainment or the expense, throughout that summer the theater's business began to slide. *Hell* may have been heavenly, but less than six months after it had opened the Folies Bergère closed its doors for good.

---

*A 1936 magazine article mentioned one more reason that the system was a dud: It blew cold air *beneath* the theater seats. "The audience objected to having this part of their anatomy cooled . . . ."

However, the Folies Bergère, and Broadway theater in general, was a high-priced playground, one that was ignored by less-moneyed people. Lasky, along with a huge number of Americans, would discover a more popular, and much more profitable, branch of show business. It had actually made its bow some years before, as part of an 1896 vaudeville show, at Koster & Bial's Music Hall, which had long boasted its own cooling machine—Thomas Edison's "Vitascope."

Movies.

The day after that first Vitascope showing, Broadway producer Charles Frohman, thunderstruck, had told the *New York Times* that the possibilities of film were "illimitable." And he was right.

By the turn of the century, tiny improvised theaters began to spring up, devoted to showing the new moving pictures. It was a small investment: a commercial space, window coverings, a hand-cranked projector and someone to crank it, a white sheet for a movie screen, as many chairs as it was possible to cram into the place—and presto, a greengrocer or a hardware man or even a funeral parlor manager was suddenly An Exhibitor. Even better, the standard admission price was five cents, one-fifth that of the least expensive Broadway ticket, affordable by nearly anyone. In 1905 a Pittsburgh theater operator capitalized on the price when he copped the name of a Boston music hall, The Nickelodeon, for his own moving-picture house. The name instantly swept the country. Nickelodeons became an unprecedented craze, popping up everywhere; by 1908, one estimate counted 8,000 of them nationwide. Millions of Americans had their first taste of the performing arts as "nickelodeon fiends": Kids playing hooky, unchaperoned women, men returning from work, courting couples, even entire families dropped in for an hour's cheap entertainment, sometimes two or three times in a single day.

But as entertainment experiences went, the fun provided by nickelodeons could only be called . . . rigorous. Being fashioned out of anything from storefronts to stables, plenty of them lacked basic sanitation. Seating was often catch-as-catch-can. And most of the time, no ventilation was provided at all. Interiors needed to be pitch-dark, so windows were heavily covered to the point of airlessness. If air came in, it came in through a door that might be cracked open. One historian remembered nickelodeons as "hot, stuffy, uncomfortable, and packed all day long." "Hot" was an understatement. Even in winter, the heat of projector lights, combined with the body heat generated by a whole day's worth of elbow-to-elbow crowds, created an unbreathably thick atmosphere that often caused patrons to keel over in the middle of a film.

Harper's Weekly disapproved thoroughly of what it called "Nickel Madness." This drawing of an airless, crowded nickelodeon ran with an article about the movies and fittingly enough was captioned "A Narcotic Quality Pervades the Warm, Fetid Air." (Library of Congress, Prints and Photographs Division)

But that word "stuffy"—it was a clue to an issue that couldn't be solved merely by opening a window. Even though nickelodeons were unbelievably profitable, they were dismissed as a type of amusement suitable only to The Lower Class: While some of that reputation came from the rough-edged films themselves, and some from the fact that men felt free to spit on the floor and women to nurse their infants, most of that reputation came from the smell. Reports flooded in from everywhere, sneering at "nickelodeon air" and its assault upon noses: "Inside a theater one's first impression is of stale, still air and the smell of sweat and unwashed bodies"; "they smell of cheese and garlic"; "the smell of body odor, cheap perfumes, powders and liniments"; "The air in most of the movies is unconscionable. That is putting it mildly."

At first, nickelodeon owners tried to cover up the problem. Literally. A Pittsburgh exhibitors' trade paper ran an ad for Denol Germicide: "Sprayed throughout your house [it] will kill all germs and at the same time impart a delicate perfume." If an exhibitor couldn't find Denol, there was a mix-it-yourself recipe published in *The Druggists Circular* for a "perfume solution—not a perfumed deodorizer—to be used by evaporation within a nickelodeon theater." As to the results, a group of New York inspectors reported with as much amusement as horror, "Perfumed disinfectants are being sprayed in some of the theatres with a cattle sprayer. The device was of the variety used in killing fleas on the bodies of animals. In some places when a baby with the whooping cough attracted attention the attendant would walk over, squirt the deodorant upon it, and permit it to remain where it was." The report didn't give the babies' reaction to this treatment.

If some exhibitors thought that a squirt of scent could solve everything, others understood that moving pictures weren't going to become attractive to higher-income audiences until their playhouses became more inviting as well. For a while some exhibitors resorted to the airdome, a fanciful name for a theater without a roof. A minority of these were solid, comfortable buildings with complete stage equipment, but the usual airdome was described spitefully by *The Green Book Magazine* as "a large or small vacant void surrounded by a knothole-proof enclosure." Most were no more than an empty lot enclosed by canvas barriers, offering long benches for seating, spartan to nonexistent sanitary facilities, and a flapping cloth movie screen. Their only upside was the promise of air, plenty of it, an unobstructed view of the heavens, and summertime relief from the heat. Their problem was the same one that afflicted roof gardens—if it was a steamy night, or if it rained, audiences were out of luck.

By the 1910s, the moving-picture industry was expanding as the old single-reel "flickers" were starting to give way to more polished, feature-length entertainment—a bigger expenditure, meaning that it was becoming urgent to broaden the audience base with more middle-and upper-class patrons. In response, exhibitors across the United States were replacing their nickelodeons with larger theaters, designed specifically for movies. And while the book *Picture Theatre Advertising* gave a whole list of suggestions for hot-weather business, one of which was that exhibitors

Airdomes could advertise better air than nickelodeons, but they had their own problems; shows couldn't start until the sun went down, rain would kill the entire night's business, and there was the constant threat of mosquitoes. (Courtesy of the Saugatuck-Douglas Historical Society)

could simply bluff the public—"[Persuade] the patron and the passerby that your house is cooler than any other place in town. . . . Tell the entire staff that you'll fire the first person who uses a palm leaf fan or admits that it is a hot day"—others tried to provide some kind of ventilation. Some of those attempts were intriguing, if inadequate. A Savannah theater was "built out into the water, which makes it the coolest place of any moving picture theater in or around Savannah." A Houston theater was built with "forty large openings in the walls" to admit outside air. The Star Theatre of Smethport, Pennsylvania, gained a door "cut through the east wall . . . which will add to the ventilation as well as make a suitable fire escape."

Most theaters just gave in and installed fan systems. The Pictorial Theatre in Hillsboro, Texas, was sold to new owners who installed "a perfect system for proper ventilation, new ceiling fans and more wall fans"; to underline the renovation, it was renamed the Wonderland. In Panguitch, Utah, the Star Theatre was renovated and renamed the Elite. Part of the improvement was three electric fans "to keep the theater properly ventilated and sanitary."

The promise and the reality. *Above*, the Comet Theatre advertises itself as the COOLEST THEATRE IN NEW YORK and promises ICED AIR; while *below*, an unidentified movie house is short on glamour—including its ventilation system, which consists of two electric fans. (New York Public Library, Picture Collection; Library of Congress, Prints and Photographs Division)

Larger cities upped the ante, adding splendor to the experience. New York opened "Roxy" Rothafel's Regent in 1913 and the 3,000-seat Strand a year later, the first-ever bona fide movie palaces; two years later came the even more deluxe Rivoli, flagship of the Paramount Pictures chain. The Rivoli wowed not only moviegoers but also *Motion Pictures Magazine*, which described one of the theater's many luxuries: "a novel feature . . . 'olfactory music,' or a system of atomizers which spray perfume—oriental, clover, new-mown hay—and in accord with the orchestra, the screen and the stage settings literally imbue one's senses with the atmosphere of the play." The magazine may have called it music, but everyone else understood that it was nothing more than an upscale version of that nickelodeon attendant wielding his cattle sprayer.

At the time, there were two choices for theatrical ventilation: an exhaust system, which pulled hot air out of the theater (and required, for fresh air to enter, openings that were rarely provided, which meant that a theater's front doors would usually be left open, which meant that the last row of the audience was sure to be annoyed by air blowing steadily on their necks throughout the show), or a "plenum" system, which reversed the whole thing by forcing fresh air into the space and (theoretically) forcing out stale hot air. Proponents of each system accused the other of being costlier and noisier. The Typhoon Fan Company specialized in the plenum system, and two Typhoon engineers broke away to offer their own aptly named Monsoon Cooling System: "Large quantities of fresh air blown into the house, creating pleasant, cooling breezes, evenly distributed." Both the Strand and the Rivoli had been outfitted with Typhoon installations. *Architecture and Building* admired the "mammoth" Rivoli system: "The air passes out under its own pressure through the front doors and through openings above the rear of the balcony."

But there was an issue with such systems, and it had to do with the way the air came in. The simplest option was to install fans on the roof and have them blow into the theater, in which case some patrons were sure to complain of being "blown on." A more elaborate system would bring in air under the floor through "mushroom" ventilators, which looked like their namesake, placed beneath every few seats. There were problems here, too: If the air came in too slowly, patrons got little relief from the heat. If it came in with more strength, managers would get outraged complaints, particularly from women who claimed that air was being blown up their skirts. And as hot, stale air rose, the balcony would become more and more uncomfortable unless an exhaust fan was in place to pull out the heat. But whether the air was pulled in or pushed out,

This slide was projected in theaters as an attempt to convince patrons that they were comfortable. (Reproduced from *Perils of Moviegoing in America: 1896–1950*, Gary D. Rhodes, The Continuum International Publishing Group, 2011)

moving quickly or slowly, there was a more basic problem with fan ventilation—if the "fresh" air was hot, so was the audience. The Monsoon System might claim that it rescued summer business, but the truth was less happy; with or without any kind of ventilation system, only the most diehard moviegoers would frequent movie houses in the summertime. The studios recognized this and cut their losses by releasing only quick-and-cheap films until autumn.

Very grudgingly, exhibitors began to look at ways to cool the air in movie houses. Part of their reluctance was the fact that ammonia, as a refrigerant, was unpleasant stuff. Should a leak develop, and the odds were pretty high that a system would develop at least one leak during its life, ammonia fumes would instantly make their way into the cooled space; the ensuing panic would be, at the very least, bad for public relations. (Even the installation at the New York Stock Exchange was fated to experience an ammonia leak, in 1924. Fortunately it would happen before the start of that day's trading, but the *Washington Post* made sure to note that "when the gong rang, slight traces of the fumes were in evidence.") So, many of the first attempts at movie-house cooling tried to circumvent the issue by using ammonia-free air washers. To provide more comfort, some exhibitors slapped together low-tech solutions that seemed like throwbacks to the cooling-machine era: The Isis Theatre in Houston installed an air washer as part of its ventilation plant as early as 1912 and tried plopping blocks of ice into the water tank to cool things down. Plenty of other theaters, among them super-luxury houses such as Sid Grauman's Million Dollar Theatre in Los Angeles, followed suit, but the result couldn't be controlled. Other uses of ice were even less scientific. Chicago exhibitor Barney Balaban tried to cool a theater by using fans to blow air over ice. As his nephew Bob Balaban later related, "In the beginning it worked pretty well, but [the fans] were very, very noisy, and also occasionally the ice would start to melt, the fan would get out of control, and all the patrons would be splashed with icy cold water."

Then higher technology got involved. In late 1917, the New Empire Theatre in Montgomery, Alabama, installed "one 6-ton vertical single-acting belt driven enclosed type refrigerating machine," making it the Very First Air-Conditioned Movie Theater. *Motion Picture News* called it "One of the South's Finest Theatres," but there was little reaction in the popular press—possibly because Montgomery was off the beaten track, more likely because the system was seriously undersized for its space.

But only weeks later, Barney Balaban opened the Central Park Theater in Chicago with partner Sam Katz and elevated the movie palace to previously unknown heights of fantasy fulfillment, all of it available to anyone for the admission price of ten cents. Every public space, from lobby to auditorium to restrooms, was a study in extravagant voluptuousness. Each patron, rich or poor, was unfailingly treated as royalty. Ushers—an ad promised, "These ushers are *picked men*"—were drilled into abject politeness by a West Point graduate. And, the ice-and-fans contraption having been abandoned forever, the Central Park was thoroughly, unmistakably . . . *refrigerated.*

This came about through serendipity: When Balaban hadn't been managing theaters, he had been working in the meat-packing trade at the Western Cold Storage Company and getting a daily demonstration of mechanical cooling. At the same time, engineer Frederick Wittenmeier, working for Kroeschell Brothers Ice Machine Company, had been perfecting a refrigeration system that was powerful enough to cool large rooms while using a (comparatively) nontoxic substitute for ammonia, carbon dioxide, as the refrigerant. It required machinery that was much bulkier than anything used in an ammonia plant, but for a movie palace, that wouldn't be a problem. Balaban met Wittenmeier and saw the possibilities; one historic contract and 15,000 feet of refrigeration piping later, the Central Park Theater was equipped with a massive Kroeschell system, filling two basement rooms, that could maintain an indoor temperature of 78 degrees even when the outside thermometer registered 96. It was a sensation. And a profitable one. Within a few years, there were more Balaban and Katz theaters, and one of their trademarks was that they were *refrigerated*. In the summer of 1919, B&K's Central Park and Riviera theaters were trumpeted in one of moviedom's earliest icicles-hanging-over-the-words newspaper ads. Obviously air conditioning had become a drawing card of its own.

If there was a downside, however, it had to do with an updraft. Wittenmeier's refrigeration was delivered to theaters through the customary floor-mounted mushroom ventilators, on the theory that the air would enter the auditorium very cool, warm to a comfortable temperature, rise to envelop the patrons, and finally be exhausted through the ceiling. In practice, though, movie patrons received a blast of disturbingly cold air at floor level, most of which stayed there. More than one report told of patrons' wrapping their feet in newspapers in order to get through a movie. As well, the system concentrated on temperature control and was less concerned with humidity. The Central Park and Riviera were definitely cool spots in the heat of a Chicago summer, but there were complaints that they sometimes felt, well, *clammy*.

Around this time, Willis Carrier had entered the movie-palace business and was taking note of the various successes—and failures, some of which his own firm experienced in theater installations by following Wittenmeier's pattern. Obviously something different would be needed. Carrier engineer L. L. Lewis shocked the industry, devising a system that was called "upside down" because it delivered cool air through the ceiling and allowed it to settle onto the audience. There was some initial opposition, most of it from theater designers who felt that the usual ceiling-mounted Carrier air diffuser cones looked like what they were, factory equipment, and these designers would *absolutely not allow them to be seen* in their aesthetic conceptions; they relented only once the company came up with an artistically acceptable substitute that could unobtrusively blend in with the surroundings. Combined with the company's

This blurb was significant, if only because it wasn't touting any movie at all but praising the miraculous *coolth* to be found in both theaters. It was also notable for the fact that it took up an entire half-page of the paper.
(*Chicago Daily Tribune*)

attention to humidity control, Lewis's system gradually became the industry standard.

There still were at least two more obstacles to overcome: the cost and size of the equipment. It was reasonable for a multimillion-dollar movie palace, seating thousands, to install $50,000 worth of air conditioning equipment that would take up half the basement. But few exhibitors could operate on that scale. And still, there was the ammonia problem.

Carrier solved the problem when he developed the "centrifugal compressor"—smaller, cheaper, and, best of all, ammonia-free, using the new nonflammable refrigerant dielene, which had been around for some time but used only as a dry-cleaning fluid. *Ice and Refrigeration* reported on the centrifugal compressor's unveiling to a group of engineers in 1922, an event taking place in the Carrier sheet metal plant, at "an elaborate dinner served in the sheet metal shop, by the feminine members of the Carrier organization." To add to the men's-club appeal, guests were treated to two amateur boxing matches. And for technical interest, the entire evening was kept cool by the new compressor, set up in the next room.

While that original-model compressor would go to Syracuse, New York, serving the Onondaga Pottery Company (and at the end of its working life, to retirement-with-honor and permanent exhibit at the Smithsonian), one of its descendants would make the Hollywood bigtime when Carrier was asked to air-condition New York's Rivoli in 1925. This was an extremely high-visibility installation, and Carrier himself supervised the whole thing. There was some difficulty at the last moment—not with the installation but with the New York City building inspector, who refused to issue a permit for the system on the grounds that he had never heard of any such thing as dielene. Carrier argued that it was being used with complete safety in a number of installations. The man remained adamant. Finally Carrier proved his point by whipping out a bottle of the stuff, sloshing it into a cup, and dropping in a lighted match. As promised, the dielene didn't explode; the inspector, unnerved by the demonstration, did. But once he recovered his temper, he had to admit that it was indeed nonflammable. And he issued the permit.

The newly cooled Rivoli was scheduled to open to the public on Memorial Day, with the words "REFRIGERATING PLANT" blazing on the marquee. It would be an acid test. The day had been hot, the system was started late, and the auditorium was still muggy when a standing-room-only crowd, "seven deep in the back of the theater," piled in. "They were not only curious, but skeptical—all of the women and some of the men had fans," recalled Carrier. More unnerving, one of the audience members

Memorial Day, 1925, the Rivoli Theatre. Both the lighted marquee and the "COOLED BY REFRIGERATION" banner introduce the theater's brand-new cooling system. Just out of camera range is an automatic thermometer, with an accompanying sign: "FACTS! This meter registers the temperature of the Rivoli—Cool—Clean—Dry Air—Comfortable." After this night, Rivoli management will realize just how well the installation works when box office receipts increase by more than $5,000 per week. (Courtesy: Carrier)

was Adolph Zukor, head of Paramount Pictures. Carrier remembered that the evening started nightmarishly: "From the wings we watched in dismay as two thousand fans fluttered. . . . We felt that Mr. Zukor was watching the people instead of the picture—and saw all those waving fans! But the temperature gradually dropped. . . . One by one, the patrons dropped their fans into their laps."[1]

After the show, Carrier found Zukor in the lobby. "When he saw us, he did not wait for us to ask his opinion. He said tersely, 'Yes, the people are going to like it.'"

The installation was a triumph for the Rivoli, which for a while would remain one of only a handful of air-conditioned theaters in New York. The Capitol, which had wrestled with an air washer since 1922, announced —coincidentally, the week after the Rivoli unveiled its system—that it was installing a "400 TON COOLING PLANT: ALWAYS 70°." But far more exhibitors were adopting a wait-and-see attitude. The following year, out of an entire page of midsummer film offerings in the *New York Times*, only two ads mentioned heat relief: the Strand, touting "NEW YORK'S FINEST COOLING SYSTEM—70 DEGREES ALWAYS," and the Rivoli, showing screen siren Clara Bow's white-hot *Mantrap* while cooling things down with "REFRIGERATED MOUNTAIN OZONE—Great refrigerating machines are at work all summer purifying and reducing the temperature of the Rivoli to just the desired degree."

However, the bigger triumph was for air conditioning itself, which soon became indelibly linked in the public mind with moviegoing—the first venue in which patrons let management know, loud and clear, that comfortably cool surroundings were *absolutely required* . . . especially when the cool comfort cost the patrons nothing more than the price of a ticket. The film industry, which had been forced to operate at a loss during the summer months, became a year-round goldmine. Exhibitors quickly realized that the appalling up-front cost of summertime cooling was actually a sharp investment; the Rivoli recouped the entire $65,000 price of its Carrier installation in an astonishing three months. And theaters without cooling were doomed to extinction.

The *Chicago Daily Tribune* noted the trend: "Theaters . . . are coming to recognize quite generally that a bunch of humans is entitled to treatment as good as that usually accorded a bunch of bananas. In a few years artificial cooling of theaters will probably be as general as the artificial heating of homes." Exhibitors agreed. Whatever the cost, thousands of them stampeded to install refrigeration. Carrier alone outfitted more than 300 theaters over the next five years, while a plethora of imitators sprang up to offer their own equipment.

Strangely, while it was happening the result of all this activity was overlooked by many observers. But in fact it was a staggering development. *For the first time ever in human history*, there was a hot-weather refuge available to overheated people,* no matter their class or income level—a refuge that was easily affordable, and dependably cool.

Exhibitors, for their part, made this impossible to miss. Each summer, it became a common sight all over the United States to see the front of

---

*In truth, the phrase should be "overheated Americans," as this development occurred almost exclusively in the United States. Paris and Buenos Aires each got an air-conditioned theater in 1929; but elsewhere in the world, new cinemas from Santiago to Berlin, even the most modern, continued to make do with fan-driven "ventilation" systems . . . or none at all.

the local movie house transformed into Arctic scenery: marquee frosted over with artificial snow, signs dripping with icicles, penguins and smiling polar bears luring patrons, a banner proclaiming some version of "20 DEGREES COOLER INSIDE," and, to prove it, the most welcome cliché of all—lobby doors propped wide open to envelop passersby in a gush of cold air. The magazine *Refrigerating Engineering* called this "advertising air" and stressed that it was "extremely desirable . . . on account of its refreshing effect to those passing the theater."

The Great White (-Hot) Way

The near-universal cooling of movie theaters old and new was not only a major innovation; it was a unique marketing coup. Logically, it should have spread next to live theater, and particularly to Broadway. But Broadway wasn't inclined to make the effort.

Even when movies were just starting up, live theater had long been a booming business everywhere in America and getting better all the time. In New York, theater was prospering as never before; the very name "Broadway" had become an unmatched symbol of theatrical glamour. Obviously Broadway houses didn't need to invest in hot-weather comfort to be profitable.

So the majority of Broadway houses didn't bother. The decades-past example of the Madison Square Theatre had been long forgotten by most people, and the Acme-cooled Folies Bergère hadn't stayed in business long enough to make an impression. Charles Dillingham's 1910 Globe Theatre recycled the old idea of the topless auditorium when the *Sun* reported "a sliding roof, which will open the house to the starlight when the run of a play extends well into the summer"—a benefit that apparently was never put to use because the roof proved to be too clumsy to operate. The Winter Garden went in for fresh air in a big way, possibly because the building had started life as a horse exchange, when it installed a monstrous fan system. The management even posed cast members of *The Passing Show of 1915* in front of it for a Typhoon ad: "EACH PERFORMANCE GIANT TYPHOONS (LARGEST IN THE WORLD) HANDLE AS MUCH AIR AS IS CONTAINED IN ALL THE STREETS OF MANHATTAN." But this was unusual; only a handful of New York theaters had an actual ventilation system. Of those, a few would revert to cooling-machine days, halfheartedly blowing air over ice in an attempt to make their auditoriums more comfortable. As it was usually not enough air, or not enough ice, those efforts came up short when they were compared with even the most primitive early air conditioning. Managements stubbornly ignored that fact. The Winter Garden (after people had stopped being impressed with the size of its fan) bally-hooed its "FAMOUS ICE-COOLING PLANT NOW IN OPERATION!"; the Empire Theatre insisted that its shows were "ICE COOLED"; and the always-in-financial-trouble Century Theatre flailed about with ads that changed frequently, proclaiming "$100,000 COOLING PLANT IN OPERATION" and "COOLED BY TONS OF ICE." There was little public reaction to the Century's strenuosities, except for a few theatrical critics who groused that the "currents of cold air" had destroyed the theater's acoustics.*

Just as in the nineteenth century, a number of Broadway houses simply closed down for the summer months—even some of the biggest hits would routinely call a hiatus until September—and serious playgoers knew enough to stay away until the fall. Producers who were caught with performances in hot weather often invested in a cheaper and simpler cooling alternative to ice. This was recalled by playwright Moss Hart in

The Winter Garden's Typhoon Fan—twelve chorus girls wide.
(*Moving Picture World*)

*As late as 1961, Broadway's Longacre Theatre was advertised as "ICE-COOLED."

his memoir *Act One* when he wrote of the punishing heat to be found in theaters during the 1920s, as well as the completely underwhelming way in which it was handled:

> Two giant-sized electric fans on either side of every theatre proscenium were kept running until the house lights dimmed, and were turned on again for each intermission, but the heat generated by an audience on a hot night was still formidable. The make-up ran down the actors' faces, and the audience itself was a sea of waving programs and palm-leaf fans, the rustle of which sometimes drowned out the actors altogether.

The entire theatrical world stoically put up with the discomfort through the 1920s without a loss of business—the 1926–27 season would set an all-time record, when the city's 40-some theaters played host to a whopping 297 new productions. But things changed for the better when one of the productions, *Rio Rita*, opened the brand-new Ziegfeld Theatre, the Great Glorifier's custom-designed showplace. It was designed to provide the ultimate theatrical *and* environmental indulgence; when its plans were first unveiled, the *New York Times* had reported, "There will be a cooling plant, capable of keeping the theater at 50 degrees if desired." Carrier installed the system, and by the summer of 1927 newspaper ads announced that *Rio Rita*'s audiences would be pampered in the "COOLEST THEATRE IN THE WORLD."

That did it: Bit by bit, air conditioning began to show up on Broadway. Only three months after the Ziegfeld opened, the Keith-Albee vaudeville circuit announced that they were installing air conditioning in six of their New York theaters, including the Palace, and one in Cleveland. Later in the year, a new Hammerstein's Theatre opened with an "elaborate air conditioning system . . . one of the greatest ventilating systems ever installed in a theatre." At the end of 1932, the 6,000-seat Radio City Music Hall opened in Rockefeller Center with every amenity known to the art of stage production, including Carrier air conditioning. By 1937 the *Times* would note that, of fifteen shows still playing in late June, eleven of them were housed in air-conditioned theaters.

Caveat emptor was still the rule, however, proved by the saga of the theaters' neighbor down the street, the Metropolitan Opera House, and its attempt to update its 1880-vintage ventilation. In 1934, it announced a drive to raise funds for a number of improvements, including "installation of an air-conditioning system." That was premature; after trustees

The Casino Theatre catered to a very high-end audience, but the tiny fan mounted at the left of this theater box would have provided precious little comfort to the occupants. (Byron Company/Museum of the City of New York)

were made to realize that ductwork could be installed in the famous auditorium only by tunneling into the Met's solid brick walls, the whole idea was downgraded and an air-*circulating* system was installed instead. The quality of the system would be memorialized by a Met historian, who wrote that it "merely substituted street air of nature's own temperature for the murk of the auditorium." Even though the system would still be called "air conditioning" well into the 1940s, no one was fooled. At least not twice.

More exciting vistas had unfolded when the $4.5 million Earl Carroll Theatre opened in August 1931. Carroll, famed for his eyebrow-raising *Vanities* revues that were famous for hordes of undraped females, made sure that his spectacular new showplace exactly filled his, and their, needs. A sign over the stage door proclaimed, "Through These Portals Pass the Most Beautiful Girls in the World." Backstage, big-spending stage-door Johnnies were nicely accommodated; there was thoughtfully provided not only a refrigerator for showgirls' floral gifts but also a safe to hold any incoming baubles. And to top everything else *The New Republic* noted "a cooling system not only in the auditorium but on the stage."

Carroll's ladies might not have appreciated the onstage cooling. After all, most of them worked in the nude.

Cheap at Any Price

The showbiz world had known forever that enticing surroundings and slick entertainment were a must when it came to bringing in the crowds; adding the novelty of year-round comfort made the combination irresistible. And nontheatrical entrepreneurs were discovering that the same formula applied when it came to drawing customers.

While Americans hadn't invented the concept of shopping as a leisure-time activity, New York retailer A. T. Stewart gave it a monumental push in 1846 when he built himself a colossal new establishment on Broadway. Nothing like it had ever been seen anywhere, five elegant stories of white marble with a central rotunda, stuffed with retail temptations of all kinds and aimed directly at women with time to kill and money to spend. Even more amazing, customers were encouraged to come in and, if they liked, spend the entire day merely browsing. Enthralled shoppers fell in love with the place and christened it "The Marble Palace." The phenomenon of the American department store was born, and with it the dazzlingly materialistic culture of The Ladies Who Lunch.

Over the next half-century department stores spread over the Western world, many of them enormous fantasy settings that promised a taste of high society to anyone who might like to visit . . . and, at some point, possibly make a purchase. To entice them, there were red carpets from the curb (literally), bowing-and-scraping employees, splendiferous opulence, and the kind of perks that most customers didn't get at home: multi-mirrored lounges, live orchestras, free telephone service, elegant restaurants, ultra-grand staircases, even *elevators*. (Lord & Taylor's 1870 building had one of the very first department store elevators anywhere; so many people rushed to "have a try" that management considerately installed an upholstered divan in the cab for their comfort.) Shopping turned into a day-long amusement and an obsession for those who could afford it. The stores themselves became tourist attractions, complete with postcards: "One of the City's Sights," read a caption. One ten-block stretch of New York was so thoroughly crammed with department stores that it became known as the Ladies' Mile. As stores became more and more successful, they expanded and rebuilt, becoming ever larger.

And hotter.

At first, department stores tried to cool themselves by using the same techniques they had relied upon when they had been small shops. Arnold Constable in New York tried Gouge's Atmospheric Ventilator, guaranteed to lift hot air out of a room—with the help of "a gas jet, kerosene lamp, or other equivalent." A. T. Stewart relied on open windows and skylight ventilators. B. Altman's had window shades that "operate both ways, rolling from the centre, thus permitting ventilation by lowering the tops of the windows and raising the lower sashes." Lord & Taylor preferred a schedule, carefully opening all windows one hour before each day's start of business.

None of it really helped. The typical department store was a structure with cast-iron walls, huge windows, and a skylighted rotunda, which meant that it functioned rather like a gigantic smokestack; or it tended to be a structure that had started small and annexed neighboring buildings as the business expanded, which meant that it had no logical ventilation pattern at all. In any case, these structures tended to get hot and stay hot. Worse, open windows brought in all the dust and dirt of the street, which spelled disaster for white goods. Worse still, a hot enough day would result in a body count as customers and employees alike dropped in their tracks.* The sight of fainting ladies (and gentlemen) became a common

---

*Sadly accurate: During the infamous 1896 heat wave that engulfed much of the country, a woman collapsed from heatstroke and died while she was standing at the counter in a Washington, D.C., department store.

department store experience, and it marred their genteel manner. Their self-restraint, too; even though R. H. Macy was known as a rigid disciplinarian, for four weeks in a row during the summer of 1879 some heat-exhausted wit dared to scribble his own graffiti into the sales record: "Awful Hot."

Once it became clear that small-scale remedies wouldn't work in a large store, Macy's tried something considerably bigger in 1880 when it installed a steam-driven fan system. The extent to which the fans worked, or didn't work, might be judged by the fact that Macy's ads praised them incessantly: "The Ventilating apparatus is undoubtedly the most perfect now in use, and is the only one in operation in the United States. By a system of valves operated by compressed air forced through them by an immense air-pump, 1,200,000 cubic feet of air is changed every hour, the pure air being forced in and the foul air drawn out of the buildings . . . ." Whether or not the system was helpful, Macy's was joined by Wanamaker's Grand Depot two years later when the Philadelphia store installed its own fan system.

Over the next two decades, fan systems—steam-driven, then electric—became the norm for the well-dressed department store. But they offered little in the way of actual cooling. And they were useless in bargain basements; as they were completely unventilated spaces, the basements' heat, combined with their always-combative customers, terrified the novice saleswomen who were forced to work in them. Stores tried to fill the cool-air gap by other means, reasonable or not. *Manufacturer and Builder* claimed that the department store pneumatic-tube cash system "affords a simple and effective system of ventilation"—but no one believed it, especially the cashiers sitting in front of the tubes. Rather less hygienically, Brooklyn's Abraham & Straus offered "a barrel constantly making the rounds of the store filled with a cooling beverage," and Belk Brothers of Charlotte, North Carolina, maintained a barrel filled with ice water at their store's front entrance; five tin cups were tethered to the barrel for customer convenience. And while Paris's Galeries Lafayette would become famous for offering paper fans as a "chic" gesture, the British travel writer John Foster Fraser remembered an uncomfortable visit to New York's Siegel-Cooper, "crowded with perspiring women," and seeing a "Help Yourself" stack of fans at the front door. This was bad publicity. "The Big Store," as it was called, was the largest and most over-the-top department store in the country, with more than 600,000 square feet of extravagantly designed selling space, including a bicycle track, but its heating-and-ventilating system, designed by the redoubtable Alfred R.

The New York Wanamaker's sixteen-story Rotunda. An advertisement read, "Hot weather shopping is always comfortable at Wanamaker's. The swirl of the breezy fans greets you as you enter the door. . . . [T]he Rotunda gives perfect ventilation to all floors." But in spite of all that perfect ventilation, the store had to close early on summer afternoons.
(*New York City and the Wanamaker Store*)

Wolff, turned out to be effective only at heating. Siegel-Cooper stoutly denied there was any problem—"The store is cool and comfortable," insisted ads—but shoppers knew otherwise, and management would become desperate enough at one point to position ice blocks in front of its air outlets.

Department stores might be huge, princely, and style-setting, but by the turn of the twentieth century even the swankiest of them was getting an unlovely reputation for hot (and worse, bad-smelling) summertime air. Government representatives began to tour stores and ask saleswomen for their testimony as to the air quality in their stores, and indictments such as "unhealthy," "foul," "the ventilating system is inadequate," and "pockets of stuffy atmosphere" entered official records.

The advent of the air washer was gratefully seized upon by retailers, who loudly, and exaggeratedly, praised the healthy, fresh, *cool* air they offered. When Wanamaker's moved to New York, it purchased the A. T. Stewart store, replaced its pointless ventilators with an elaborate system that included an air washer—an Acme; the company had found a lucrative niche supplying department stores—and coquettishly informed the public that

> the windows can be kept closed. The fresh air is pumped in by pressure, being first drawn thru a regular Minnehaha sheet of falling water, which takes out every particle of dust. The foul air is withdrawn by suction. So the atmosphere is always pure, and it is always an even temperature— cool in summer and warm in winter. Stewart depended on the weather; Wanamaker depends on science.

Boston's staid Filene's had to handle two different situations: Its opulent upstairs floors, nearly deserted in summertime (the regular customers having gone to The Country), were ventilated by what was euphemistically called "natural inflow and outflow." But the always-frenzied Filene's Basement received more attention. "Perfect ventilation provides complete change of air every six minutes. The fresh air is taken from the roof washed clean of all impurities." Other department stores advertised their own air washers with exuberance, even if the language was a tad less refined: "In the Emporium department store, St. Paul, which, by the way, advertises everywhere as the fresh-air store, a wheelbarrow load of mud is taken out each week from the air washer."

Lord & Taylor moved from its outdated store to a spectacular Fifth Avenue location in 1914, having spent nearly two years teasing the public with driblets of information describing the new building's attractions: a concert hall, an open-air promenade, a gymnasium, "an entrance for customers' automobiles, leading directly into the building"—and, as to ventilation plans, "some novel features are in contemplation." The trade magazine *Domestic Engineering* described them: "There are four Carrier air washers installed to take care of this part of the building. . . . It might be added that the cooling effect from the ventilation on this store floor during some of the warm days of the summer was very noticeable. The writer on a number of occasions recalls hearing it remarked by shoppers coming in from the heated atmosphere and pavement how cool this main floor was." But as Lord & Taylor's target clientele was decidedly upper-crust, people to whom such things supposedly didn't matter, the store didn't trouble to publicize its coolness.

The fact still remained that an air washer could rarely lower the temperature of a hot building more than a few degrees. Hudson's, one of the largest stores in the United States, learned this lesson the hard way: In 1916, the retailer was jubilantly crediting the newly installed Sirocco Air Washer in its four-acre bargain basement with a $50,000 increase in sales. By the early 1920s, the honeymoon was over when it became obvious that no air washer could cope with the heat of a Detroit summer. Carrier engineer Margaret Ingels wrote—and while she was referring to Hudson's, the same story applied to department stores across the country— "[T]he ventilating system in its basement was of little help. Temperatures soared; customers fainted. The manager feared he would have to discontinue bargain sales unless he could keep temperatures down when the crowds poured in."

In 1924, Hudson's contracted with Carrier to install three of the new centrifugal compressors in its basement, a $250,000 job (more than $3 million in modern terms). But it turned out to be a farsighted investment. Ads in the *Detroit News* promised "On Warm Days It's 8 to 12 Degrees Cooler in the Basement Store than Street Temperature," and as soon as customers realized that the ad was telling the absolute truth there was an immense surge in business. Hudson's Basement Store became as popular as . . . an air-conditioned movie house. Indeed, a lady drove the point home when she wrote to the *New York Times* about a mid-July shopping trip at an unnamed store: "I saw the clerks practically exhausted with the heat before the noon hour. I was so uncomfortable I decided to postpone my shopping until there was a change in the weather, and went instead to one of the picture houses where they distribute frigid air. I found it full to capacity. When will the proprietors of stores awaken to the fact that they can make people comfortable also?"

As with movie houses, it suddenly became mandatory for any well-appointed department store to offer air conditioning *somewhere* on the premises—usually a store's basement and main floor, the areas with the highest traffic.* The first stores to take the plunge were those that could afford the enormous cost (Filene's, in 1926), or those whose climate made it crucial; when Dallas retailer Titche-Goettinger built a 1929 store with "the latest retail features for the customer's convenience," one of those features was a Carrier system.

Macy's called Carrier in 1929, too, and had fun with it by announcing the new, cool atmosphere with a series of wryly comic book–style ads. The *New York Times* ad of July 29 was entitled "Pleasant Dreams":

> Come at 9:30 and stay until 5:30, but don't ask to spend the night. We'd understand how you'd want to. We want to ourselves. It's so cool here we've been tempted many times during the hot spell to snuggle down beside a retired escalator and catch up on our sleep. But it can't be arranged: There wouldn't be room for all our 10,000 employees to lie down on the Street Floor and in the Basement, and the ones who couldn't find accommodations would get high blood pressure out of sheer envy. . . . We just have to remember that in a few hours morning will come and bring with it another cool Macy day.[7]

Air conditioning became such a draw that, no matter the cost, department stores found it profitable to install it, and stores that had it in the basement made sure to install it on other floors. Hudson's was a perfect example: Soon after the basement system was operating, Carrier was called back to air-condition the first three floors. Within a decade, the entire store was air-conditioned.

And as to whether ladies of the *bon ton* were still immune to the lure of summertime cooling, there was this ad from 1936, directed to those

---

*Some observers have pointed to this as retail management's calculated manipulation of class distinctions: that well-heeled ladies who frequented the upper floors wouldn't particularly care about temperature control and wouldn't even be around during the hot months, while working-class women who haunted the bargain basements would be impressed by air conditioning, and so forth. Others have simply assumed that the stores were trying to deal with the immense cost, and frightening amount of ductwork, involved with air conditioning by getting into it gradually, starting with their most heavily trafficked floors.

matrons who were suffering from The Servant Problem as well as from the heat:

> There are three smart ways of keeping cool on the cook's night out. . . .
> One good method is to patronize an air-cooled movie. Another is to visit
> an attractive roof—the St. Regis, for example. The third is sound and
> popular, too—SHOP AT MACY'S. For Macy's is open till 9 on Thursday eve-
> nings, and there are five whole air-cooled floors . . . .[2]

How Hot Is Soundproof?

At the same time that air conditioning was selling itself to the general public, it gained an extra sheen of glamour, and a great deal of respect, when it came to the rescue of a couple of new high-technology entertainment industries.

Take Hollywood. Movie palaces might be 70 degrees in any weather, but the movie-making process itself was considerably less comfortable. Mech, the Carrier mascot, had seen this first-hand in one of his early cartoons when he visited a movie studio. With studio brass in the background complaining about film stock becoming unworkable in hot weather, a megaphone-wielding director and a wilted actress, both frantic with the heat, burst in on the scene:

> "And how in blazes can we work"—a crazed director cries,
> "In ninety-six degrees of heat, with dust clouds in our eyes?"

If it was Carrier's subtle hint to the movie colony, it was ignored. Air conditioning had made it into the movie business as far back as 1908, but it was found only in the *labs*, having been installed to help film dry without spotting. Actors and directors had a tougher time of it.

Electric lighting was costly, so from the earliest silent-film days studio stages were built with glass roofs to catch every bit of available sun. This, of course, turned the stages into ovens. The 1921 book *The Film Industry* had recommended that "the interior should be well ventilated by electric fans, because of the heat generated from the sun shining through glass." If fans weren't being used, the roof panes might be opened to let in some air, in which case inclement weather would enter, too. And this meant more than mere rain: British studios were constantly bothered by the infamous London fog, billowing in through open skylights and visible enough to ruin shots. In a single year, shooting at one English studio had to be completely shut down for twenty days because of the fog-on-film problem.

In 1926, the movie world changed forever when Vitaphone—talking pictures—made its bow at New York's Warners' Theatre. The premiere took place in sticky August heat, something that the audience could laugh off; hanging from the theater marquee was the (by now) familiar and reassuring icicle-bedecked sign, "REFRIGERATED WASHED-AIR COOLING SYSTEM." However, the process of *making* the talkies that were shown that night was a much steamier experience than silent films had ever been.

Stages now had to be completely soundproof, and they were rebuilt to be as "airtight as a refrigerator" with concrete walls, bank-vault doors, and solid ceilings: no more open skylights. Electric fans were barred from the stages: too noisy. And the sound of the loudly clicking camera was muffled by enclosing it, along with the cameraman, in a claustrophobic booth with eight-inch-thick walls, a double-pane window through which the camera could focus, and no ventilation whatsoever. Plenty of magazines such as *Popular Science* were nervously fascinated by the booth— "the operators, sealed in soundless tombs"—but not one mentioned that the cameramen themselves dreaded it. By the end of each ten-minute reel of film most of them would burst out of the booth, gasping for air and sopping wet; some others would have to be pulled out, unconscious. They wound up giving it darkly comic nicknames of their own, such as "The Icebox," "The Tank," and "The Coffin." (Sound insulation for cameras wouldn't make it to the studios until the early 1930s.) Things were even hotter in front of the camera, as the silent-era carbon arc lights made sputtering noises and had to be replaced with incandescent lighting, quiet but so hot that makeup melted on performers' faces faster than it had in any Broadway theater. Worst of all, full-color sequences became a craze in early talkies, and color film required even more brutally torrid lighting. During the filming of one Technicolor musical number, time-out had to be called when an actor's hair began to smoke.

Producers quickly realized that totally professional movie-making couldn't happen when everyone involved was laboring under severe heat stress. The *Washington Post* noted that sound stages "become exceedingly close and warm when the doors are locked, and between scenes the players rush for the open air. This naturally has produced many colds in the head and a tendency among the afflicted actors to speak their lines with a 'sprig has cub' accent." MGM was one of the first studios to respond to the problem, installing air conditioning on a sound stage in 1927. And the following year, when the Fox Film Corporation plunged $10,000,000

The cameraman's expression says it all. (Ron Hutchinson/
The Vitaphone Project)

into a "gigantic concrete sound-proof city" of new sound stages, the *Los Angeles Times* described their ventilation with Hollywood hype:

> The new sound-proof units have a complete air conditioning plant of their own which is the largest and most complete on the Coast. . . . Great amounts of air are taken in and passed either through a sheet of water or across heating devices, depending on whether it is desired to cool or heat the air, and then passed through a long underground tunnel to the stages where the fresh air, at any given temperature, goes in at one end and out the other and then is again conducted underground and out into the atmosphere, without the aid of fans or other apparatus which might cause vibration.[3]

But many studios just did without. Even in the mid-1930s, a *New York Times* article would describe Fay Wray in the midst of a midsummer shoot at a noncooled studio, "carrying a chamois powder cloth and using it every other minute." And even though the newest cooling equipment might be virtually silent, sound technicians still fretted that it had the potential to ruin any soundtrack. They compromised by cooling the stage *between* takes, shutting off the air conditioning every time the camera rolled.

Hollywood may have equivocated when it came to installing air conditioning, but another miraculous form of electronic entertainment had no choice.

Radio was the decade's ultimate success story. When the first station was licensed in 1920, broadcasting was represented by inventor-types in rooftop shacks and hobbyists in garages crammed with odd electronic gadgets (one report called them "a raggle-taggle mob of free enterprisers"), settling down every so often to transmit "radio telephone" programs of talk, Victrola records, or amateur singing to anyone who might happen to tune in. Less than three years later it had burst into a white-hot national craze and a business on the verge of international success. It had also outgrown the garages. Just as feeble signals had been replaced by powerful transmitters, and volunteer talent was giving way to professionals, station economics had changed, too; radio's bills were now being paid by commercial advertisers, who wanted to feel that their money wasn't going down the drain. It was important that broadcasting be housed in surroundings that were reassuringly professional. Also, because jerry-rigged telephone mouthpieces had been replaced by sensitive-but-fickle "microfones," those surroundings needed to be soundproof.

There was a mad scramble for studio space, most stations winding up in converted office buildings or concert halls. New York's WJZ transmitted from several Midtown locations, including Aeolian Hall and a room in the Waldorf-Astoria. Reno's KDZK used the town's Majestic Theatre; Philadelphia's WOO operated out of Wanamaker's department store. In London, the BBC also started in retail space atop Selfridges before moving to statelier quarters. On both sides of the Atlantic, broadcasting was thrilling . . . and stuffy. Summer temperatures, combined with heat generated by each studio's electronics—along with the fact that studios cut down reverberation by being completely curtained, walls and ceilings—made them noxious places. Guest performers would do anything to avoid scheduling appearances during the summer. Employees were stuck.

In the beginning, studio personnel tried opening windows, if they had them, but that practice quickly ended. The BBC studios were very near the Thames River, and listeners were distracted by tugboat horns. New York's WEAF had to be equally careful; a musician once furtively cracked

The dawn of radio at Pittsburgh's KDKA. Potted palms, but no ventilation.
(New York Public Library, Mid-Manhattan Picture Collection)

a window open during a hot-weather broadcast, and the listening audience heard not Offenbach's "Barcarolle" but the screech of a Third Avenue Elevated train. Some studios just suffered with the heat, while others sought a remedy. The first solutions, though, were slightly unappetizing. The British magazine *Wireless World* wrote of the BBC, "Each of the present studios at Savoy Hill is fitted with an air shaft; but the studio temperature after an orchestra has been in occupation for about twenty minutes suggests that somebody corpulent must have mistaken the shaft for an exit. . . . The only method of securing ventilation which has proved satisfactory is to change studios as often as possible." Sure enough, Savoy Hill operated out of four studios, rotated in use so that they could be thoroughly aired out between programs.

In early 1923, WEAF threw a grand reception as it moved from temporary space into a princely new home in the Lower Broadway AT&T Building: two studios, along with a separate announcing booth. It was innovative, built to order, far more sumptuous than any other radio station around, completely soundproofed—and it made a great deal out of the fact that it was scientifically, *noiselessly* ventilated. The *New York Times* wrote:

A unique feature of the new WEAF will be a ventilation system which will completely change the air in the studios within three minutes. In the majority of studios means of ventilation have not been provided for fear the room would not be sound proof. The result has been that announcers who have had to remain in the studio throughout the entire program have been subject to colds and sickness as well as some of the artists. The WEAF system of ventilation has been designed so the acoustic properties of the studios will not be affected.

Of course, changing air every three minutes, if the air being changed was hot, did little good. Graham McNamee, an announcer who came to WEAF soon after the ventilation equipment did, obviously felt that the system offered no relief whatsoever. He recalled, "The coming of summer was an annual horror to studio workers" and called the station "a sweat box . . . sizzling and broiling." Some announcers would become so over-heated that, during broadcasts, they'd quietly strip all the way down to their underwear.

NBC, the first nationwide radio network, was formed in 1926 with WEAF as its flagship station. New quarters would definitely be needed. To get more space, the network moved in 1927 to five floors of a Fifth Avenue building in Midtown; to get more comfort, they called in Carrier to design air conditioning for its eight studios, control room, and reception rooms. Impressed as listeners and onlookers might be, there was more to come; at nearly the same time that NBC was moving into its new home, CBS was being formed. When it found its own five floors in a new Madison Avenue building in 1929, it too contacted Carrier to cool its studio space.* Even the BBC would follow suit when it built its ultra-modern Broadcasting House facilities in 1931 and equipped them with a Carrier system. *The Gramophone*'s Herman Klein was impressed: "While the silence can be actually oppressive, the atmosphere never can be so; for the ventilation is such that fresh air greets you everywhere from the sub-basement to the eighth floor above street level."

---

*This system served CBS well, in more than one way. When the "Buck Rogers" radio serial was being planned in 1932, sound effects men were trying to find a way to create the sound of a spaceship. Someone recalled that the ductwork in a particular studio had been causing a whooshing noise. While most shows had to work around the annoyance, the *Buck Rogers* crew installed a microphone directly in the duct; whenever Buck's ship was supposed to be traveling, the mike was opened. The result was exactly the sound any self-respecting spaceship would make.

Gaining two radio networks in short order was amazing fun for most Americans, and it had generated reams of publicity. Carrier provided some of its own when it ran a series of full-page ads in business publications. The ad in the January 1930 issue of *B'nai B'rith Magazine* was devoted to the radio industry and featured a photo of a hit NBC show in mid-broadcast along with colorful copy:

HOW DO BROADCASTERS BREATHE IN ROOMS WITH NO WINDOWS?

Walls as thick as prison walls. Rooms as sound-proof as a dungeon. How can your favorite orchestra broadcast pleasure under conditions so unpleasant?

It was pretty difficult two or three years ago. Then the artists sat in shirt sleeves, perspiration rolling down their faces.

Today your favorite entertainers are in evening dress or actually in costume. The studio still has no windows, but the air is pure and fresh, the temperature and humidity are perfect for comfort. The magic of the radio has been supplemented by the magic of Manufactured Weather— the Carrier name for scientific Air Conditioning.[4]

That magazine photo featured banjo player Harry Reser, who had definitely been familiar with the sensation of perspiration rolling down his face during a broadcast. In 1925 WEAF had asked him to form a dance band as the centerpiece of a music show sponsored by Clicquot Club Ginger Ale, whose trademark was an Eskimo boy on each bottle. Typical of the period, the show was named "The Clicquot Club Eskimos," a fun excuse to offer 1920s jazz standards with an "Arctic" theme. Reser and his men would be teamed with a mechanical husky dog that could bark on cue, and they'd perform, at least when cameras were around, in fur parkas. The show was a runaway hit with listeners; but in the WEAF studios, it was miserably hot business. By the time the Eskimos moved to their new home at NBC, they and the ersatz dog were finally able to make music in air-conditioned comfort.

Ironically, they were victims of their own popularity. It soon became obvious that NBC's eight studios weren't big enough to house shows that played to live audiences, and the network solved the problem by renting Florenz Ziegfeld's 600-seat New Amsterdam Roof Theatre—now rechristened the Times Square Studio—as a broadcast space. However, the wisdom of the time dictated that *no* radio listener should hear a broadcast that included applause and laughter coming from a studio audience. So someone came up with the idea of equipping the studio with a "glass

Harry Reser and the Clicquot Club Eskimos, comfortably in costume for
NBC (and Carrier). (Courtesy: Carrier)

curtain"—literally a stage-sized window of glass and steel, completely
soundproof, and weighing six tons. Network executives thought it was
the perfect solution; spectators would be able to see the performers and
hear the show (if only through loudspeakers), but unwanted audience
noise would be totally shielded from the radio microphones. Advertisers
were impressed, too. One commentator noted that nearly every sponsor

using the Times Square Studio mentioned the glass curtain on the air. And particularly its weight.

But whatever ventilation the audience enjoyed, it didn't reach performers on the other side of the glass curtain. Harry Reser and the Clicquot Club Eskimos, as one of NBC's top attractions, now had to do their broadcast from the Times Square Studio each Friday evening—dog, parkas, and all, blazingly illuminated by the stage lights no matter how hot the weather. Fans loved to see the Eskimos as well as hear them; but at least one colleague remembered their nearly toppling with the heat, and as late as a 1959 interview Reser recalled those Fridays with a polite shudder.

NBC realized that they had completely underestimated their needs and quickly made plans to move house. On a rainy night in late 1933, the network kicked off broadcast operations in a gigantic studio complex in the new RCA Building: *Time* noted thirty-five studios, beyond-luxurious surroundings, and what was billed as "the biggest air-conditioning job ever attempted." Carrier installed sixty-four compressors as the basis of an unprecedented million-dollar system that was guaranteed to be broadcast-quality quiet. *Popular Mechanics* noted, "Ventilation ducts were

NBC's "glass curtain," guaranteeing that the home audience wouldn't hear the studio audience. Also guaranteeing that performers would be sautéed by the stage lighting. (*What's On the Air*)

equipped with mats of fireproofed seaweed to filter out noise." The system's "stacks" would be visible to anyone visiting the rooftop RCA Observatory, but they were made obtrusively unobtrusive by being "painted in bright colors to give the appearance of a ship's deck."

Not only was the system a hit with everyone who worked for NBC, but its huge, handsomely designed, dramatic "control panel" was used as one of the photos that promoted studio tours.

Washington's Hot Air (Part V)

It seemed that nearly everyone in the United States was discovering the hot-weather relief of air conditioning. But Washington, proud of being a law unto itself—and filled with lawmakers who were desperately frightened of appearing soft, or possibly un-American, by giving themselves some pricey new-style cooling machinery—did its best to ignore it. This attitude, combined with the climate, might have been the reason that some foreign governments had long classified Washington as "subtropical" and "a hardship post."

This photo showed only some of the sixty-four dials that monitored air quality and (according to the trade journal *Inco*) "[provided] refrigeration equivalent to that of the entire summer ice consumption of a city the size of Dallas." (New York Public Library, Mid-Manhattan Picture Collection)

After Woodrow Wilson had ousted the ice-and-fan system that cooled the Oval Office, nothing else happened to give the President, or any other lawmakers, relief from Washington summers. They coped with the heat in much the same way they always had, leaving town whenever they could and switching to white linen suits and Panama hats when they couldn't; and every May, the start of the warm season was acknowledged when Charles Thomas of Colorado would enter the Senate chamber minus his toupee (an occurrence that invariably made nationwide news).

They also took refuge in bitter complaints about the ventilation. In 1924, there was once again a "study" to test the wholesomeness of interior air in the Senate chamber—once again, amid vows that the air was perfectly pure. Senator Royal S. Copeland, who had started his own career as a physician, disagreed violently; as *Time* wrote, Copeland reminded his colleagues that thirty-four senators had died "in harness" over the previous twelve years. "Probably every one of them had his life shortened by the frightful conditions of the chamber. . . . [T]his thermos bottle is still doing deadly work to the health, strength and comfort of Senators." Later that year, the old proposal was resurrected to give the Senate chamber windows by breaking through to an outside wall. The idea would be debated, in excruciating detail, for nearly four years. By early 1928, House members had picked up the same idea for their own chamber. But at the same time, a commission of "public health experts" had begun to investigate the possibility of installing air conditioning in the Capitol.

There was resistance from some lawmakers, who were suspicious of yet one more "mechanical" fix. As well, there were those who wondered, publicly and on the record, if a newfangled invention like air conditioning might make Congress seem as frivolous as . . . as a moving-picture theater. Commission members counterpunched by stating that proper ventilation would provide not only comfort but also health benefits. And to illustrate that fact, they ominously pointed out that, in the previous 35 years, nearly 300 congressional members had died while in office. That statistic was enough to frighten lawmakers into action: Five days after the outside-wall plan had been approved, it was "indefinitely postponed" and Congress voted $323,000 for air conditioning to be installed in both the House and Senate chambers.

Carrier was hired to do the job, readying the House by December 1928 and the Senate eight months later. In order to preserve the architecture of both rooms, the glass panels of the skylights were raised two inches to provide space for invisible air outlets, an example of engineering sleight-of-hand that Carrier called "beautifully designed" in a brochure, and one that made the installation so unobtrusive that it dumbfounded some members when they entered their respective chambers to find, apparently, nothing changed but the air quality. Senators got a notice explaining the matter:

> The sensation of chill experienced upon entering the Senate Chamber is due principally to the dryness of the air causing the evaporation of the slight amount of moisture of the skin. After the completion of this evaporation the body will be perfectly comfortable. . . . No fear may be felt by the occupants of the Senate Chamber from the conditions produced by this new system of ventilation and air conditioning.[5]

Still, the historian Marsha Ackermann pointed out that true Washington politics-as-usual were maintained on May 28, 1929, when Mississippi Democrat John E. Rankin made the first "official" complaint that the House chamber was too cool: "This is regular Republican atmosphere, and it is enough to kill anybody if it continues." Other representatives applauded.

Air conditioning might have made it to the Capitol, but there was no such luck at the White House. Calvin Coolidge, a native Vermonter, was disgusted by the Washington heat, particularly in the Oval Office, but seemed defeated when it came to coping with it. *Forbes* wrote, "As to the Executive Offices, Mr. Coolidge not only found them stuffy and muggy, but the very smell offended his nostrils. He kept on his desk a gadget filled with chemicals supposed to purify, or at least deodorize, the air."

Coolidge's successor, Herbert Hoover, was a trained engineer and not willing to stop at an Airwick. So as soon as he got settled after his inauguration, he decided to solve the hot-air problem himself. The *Los Angeles Times* reported, "After several experiments he finally had an electric fan placed on the floor under his desk. As the air close to the ground is always cooler than the air circulated at an elevation, the President decided that air currents churned upward would be more refreshing and would provide a clearer atmosphere." While "the experiment was successful to a degree," it failed entirely under the onslaught of 1929's summer heat. Hoover gave up and spent his off-hours away from Washington heat at Rapidan Camp, the presidential retreat in Virginia's Blue Ridge Mountains.

Then on Christmas Eve, a fire gutted part of the West Wing.

Less than two weeks later, the *New York Times* headlined "'Pure air' Proposed for Executive Offices" and announced that air conditioning, not an ice plant, would be installed (with its machinery "under the back stoop of the building"). "Herbert Hoover will be the most comfortable President in history, as far as temperature is concerned."

The *Times* promised that the President would be back in a rebuilt Oval Office, behind his desk, "kept cooled by an artificial device," in April— only three months after the fire.

Which also was six months after the stock market had crashed.

5   Big Ideas. Bold Concepts. Bad Timing.

Most movie houses. Some Broadway theaters. Luxury hotels. A big handful of department stores. High-visibility tie-ins with the new entertainment media. Even a bank here and there. It seemed that air conditioning had finally won some broad-based public acceptance. Even better, it had expanded into a few areas of daily life as an absolute necessity. Now it was gearing up to ride the crest of 1920s consumerism.

The attempt couldn't have come at a better time: A boom economy, with extra fuel from the income provided by fat stock dividends. Mechanized production that lowered prices and made even big-ticket items readily available. The example of radio, which had proven to manufacturers that a completely unfamiliar gadget, if it was promoted in the right way, could instantly become a fixture in the American home. A slick new advertising style dedicated to selling not only goods but also the virtues of conspicuous consumption and snob appeal, and selling them with the crafty assistance of never-before venues such as the airwaves and the silver screen. And to make it all available to every level of the public, that wildly popular new form of consumer credit, Easy Little Payments. . . . It was no coincidence that the phrase "super-salesmanship" had been popularized during this time. As well as the phrase "Keeping up with the Joneses."

Therefore it was only logical, as the technology of air conditioning was finally beginning to catch up with the public's fantasy, that the makers of cooling machinery would try to produce the next big-ticket status symbol for home use. At the same time, still other businesses saw the exciting, and very profitable, potential of bringing air conditioning to the public in a variety of different places—railroads, office buildings, even automobiles. There were fascinating vistas ahead.

And most of those plans were blown sky-high by the Great Depression.

Comfort in Your Own Home (Well, Maybe Not *Yours*)

At the beginning of February 1929, the *Atlanta Constitution* covered the Frigidaire Corporation's recent convention, where sales figures unveiled the exciting news that Frigidaires of all sizes had been overwhelming favorites of the buying public. But that wasn't all: New products were announced, among them "an electric room cooler operated on the electric refrigeration principle."

This was a shrewd move, or seemed to be one. Starting in the late 1910s, "automatic" refrigerators had begun to show up in a tiny minority of households. At first they were little more than clumsily converted ice-boxes, bristling with exposed motor works and extra piping; ugly, noisy, and undependable things, they dripped oil, frightened housewives, carried astronomically high price tags (one model came in at $700, the equivalent of more than $7,500 in the present day), and in a few instances developed refrigerant leaks whose fumes killed everyone in the house. But they slowly gained in reliability and dropped in price until, by the end of the 1920s, they had become steady, affordable, and hugely popular fixtures, a must for any "modern"—in those days, a synonym for "trendy"—kitchen. Manufacturers were so encouraged by that success that it appeared to be only a simple step to begin with the idea of the refrigerator's cold-making machinery, retool it into a device that could cool an entire room . . . and, most important of all, convince the world that it was a must-have.

Frigidaire, which had built itself into the industry's front-runner, was the first to give it a shot. Starting in the spring of 1929, visitors to any of the nation's luxurious Frigidaire Showrooms could see not only refrigerators of all sizes and designs but also the newest member of the family. And they could read all about it in a brochure:

> FOR GREATER COMFORT IN HOT WEATHER—THE NEW ROOM COOLER
> It is simple, both in design and operation. A specially designed fan draws air in at the bottom, passes it over a large Frigidaire cooling coil, and then forces it out into the room at the rate of 450 cubic feet a minute. In this way the air, to a height of several feet above the floor, is cooled, dried, and circulated at a rate which produces the best cooling effect without annoying drafts or breezes.

The brochure danced lightly over the fact that the new invention was a "split-system" unit, meaning that the cooler unit would have to be connected, by means of piping, to a separate compressor ("may be located in the basement, or in any other convenient location"). And the Room Cooler couldn't be stored away at the end of the summer season, as merely a summertime addition to a home, considering the walls and floors that would have to be ripped up in order to lay the necessary piping—let alone the fact that the cooler unit alone weighed nearly 200 pounds, and the compressor an additional 400.

While even its most enthusiastic admirers admitted that the Room Cooler was capable of lowering the temperature of a room by no more than ten degrees or so, it was seen as a revelation; members of the public had never before been able to buy a machine that could actually produce cool air in hot weather. The Room Cooler snagged a lot of newspaper space. It was a proud feature not only of Frigidaire Showrooms but also of the company's exhibit at the 1929 Barcelona World's Fair. And *The Outlook*, which published a weekly column of new household curiosities ranging from improved shower caps and fireplace tiles to waterless cookers, was simply dazzled. It assured its readers that the Room Cooler not only cooled the air but also "reduces humidity and helps humanity."

In the midst of the adulation, that article happened to point up one of the Room Cooler's major drawbacks by stating, "It looks sort of like a radio cabinet"—but in truth, it didn't. Over the previous decade, radio manufacturers had learned that they could gain admission to the American living room only by hiding every bit of telltale electronic innards in a splendiferous wood cabinet. This new machine disdained that wisdom: a four-foot-tall metal shell, it looked like nothing so much as a piece of kitchen equipment and was finished in a gray enamel curiously reminiscent of the "Glacier Grey porcelain" that the company had been pushing as a decorative option on its refrigerators. While it might conceivably fit into an aggressively modernistic setting, it would provide a grotesque contrast with the Jacobean woodwork that most homeowners favored at the time.

There were a few other things that neither the press coverage nor the brochure discussed. One of them was the Room Cooler's voracious appetite for electricity; most household wiring would have to be upgraded, or a unit would draw a fuse-blowing, and probably illegal, portion of the total power supplied to the average home. And there was the price, $800 (nearly $11,000 today), same as that of a Pontiac roadster. Even in the spendthrift 1920s, the Room Cooler's first customers came solely from the ranks of high-flying businesses that didn't want to bother with ductwork (such as New York's Hollywood Restaurant, where a battery of Room Coolers provided comfort for customers as well as the mostly topless chorus line), or even higher-flying private citizens (such as National

The Frigidaire Room Cooler, which looked suspiciously like its namesake refrigerator. Catalogue copy tried to neutralize the effect by suggesting that it might be "built into a special cabinet, into different types of furniture, or placed behind grilles in the walls of rooms." (Courtesy of Scharchburg Archives, Kettering University)

Cash Register magnate Colonel Edward A. Deeds, who ordered multiple Room Coolers; they were hidden behind decorative grillwork in the staterooms of his 200-foot yacht *The Lotosland*).

Only a few months later the stock market crashed, making it imperative that Frigidaire perform a quick about-face. Over the next year, the Room Cooler was redesigned: not much prettier but far less obtrusive, enameled in "walnut lacquer on steel." And while the initial rollout had occurred without heavy advertising, in 1931 the Room Cooler was reintroduced to the public through an intensive ad campaign. Significantly, these spots showed up only in business-related magazines—*Merchandising Week*, *Time*, *The Nation's Business*—and were aimed squarely at the tax-deductible world of executives, famed not only for self-indulgence but also for a schoolboyish infatuation with expensive gadgets, and possibly the only people left in the Depression-pummeled country who had access to disposable income. Every word of the ads swaggered with Business Machismo, stressing not mere sissified comfort but something much more significant . . . the Room Cooler's all-important contribution to management efficiency: "Ready to transform hot, stuffy offices into places with an atmosphere so fresh and invigorating that 'nerve-fag' never has a chance." And the whole campaign was pegged on a catchy new phrase—actually, a catchy old phrase, the centerpiece of that now-forgotten 1901 *Chicago Tribune* article: "Turn on the Cold!"

The most unrestrained squibs ran in the *Wall Street Journal*. During July and August, a series of full-page ads appeared in the paper, each one adorned with a drawing of a high-powered businessman suffering from the heat in one way or another. On July 2, a defeated-looking executive sits at his desk under a striped umbrella, inexplicably located on the roof of a skyscraper:

> Clear thinking calls FOR COOL SURROUNDINGS
>   SO . . . TURN ON THE COLD!
>   Nobody can do a decent day's work in a hot, stuffy atmosphere—yet the work must be done! . . . And it CAN be done—in comfort—with one of the new Frigidaire Room Coolers in your office . . . ask your secretary to 'phone us today.

On July 31, a wilted executive stands slumped, reading a ticker tape:

> Quietly and gently the Frigidaire Room Cooler will absorb the hot, muggy air in your office to send it forth again cool and dry and invigorating.

On August 7, another wilted executive mops his brow in front of a wall chart:

> Your Frigidaire Room Cooler will start quietly flooding your office with cool air . . . killing the energy-sapping effects of humidity and enabling you to leave your office at the end of the day feeling fit and refreshed. . . . Ask your Secretary to get in touch with us by 'phone . . .

On August 20, yet another wilted executive sits at his desk, mopping his brow:

> Go to your office, "turn on the cold!" and make cool, refreshing weather to order . . .

And on August 31, still one more wilted executive sits at his desk, finally having come to a decision as he reaches for the 'phone:

> MEMO: Phone Frigidaire to
> TURN ON THE COLD!

Only a few days later, Frigidaire announced that its entire high-end product line, which included Room Coolers, would enjoy a 10 percent price cut. Still, having produced an estimated 4,000 units and selling only a fraction of them, the company was rumored to have sustained a $2,000,000 loss on the Room Cooler project.

Whether or not this campaign actually resulted in businessmen's secretaries' rushing to 'phone Frigidaire, other companies decided it was a good time to get into the game. General Electric came out with its own version of a room cooler in 1930; that one was a single unit, incorporating the compressor and eliminating the extra piping (but weighing more than 550 pounds), that shunned the "Glacier Grey" enamel and made some slight nod to homeowners' decorative schemes by mounting all the exposed machinery atop a carved walnut base. Attractive as this might have been,* the GE room cooler listed at $950—as much as *two* Model A Fords—and was entering the market at an abysmally bad time. During two years' production, fewer than three dozen units were ordered.

---

*Not to everyone. *Heating and Ventilating* ultimately ran a sour-toned diatribe, "Will Air Conditioning Cabinets Follow Radio's Bad Example?"

GE's entries in the room cooler market—one outfitted in Machine Aesthetic, the other disguised as a radio. (Courtesy of miSci: The Museum of Innovation and Science)

One of those room coolers was installed in the home of Willis Carrier, possibly for Carrier himself to get some sort of handle on exactly what was being foisted on the public. For nearly a quarter-century, he had been responsible for something that he called *air conditioning* and that had as part of its definition a precise method of controlling humidity. While the Frigidaire and GE units would purportedly lower humidity as excess moisture condensed on their coils, it wasn't a controlled process, and this dissatisfied Carrier.

Because of this dissatisfaction, it took somewhat longer for the Carrier Engineering Corporation to decide to enter the market with a small air conditioner of its own, the Atmospheric Cabinet. It wasn't at all perfect, another hulking "split-system" unit which also hadn't figured out how to handle the problem of precise humidity control. But it bore the Carrier name, and that was more than enough to give it a flashy launch: In mid-1931, the first six Atmospheric Cabinets were installed in the New York offices of Lehman Brothers.

Nine months later, they were officially available to the average home-owner, improved and using the brand-new refrigerant Freon. (In some ways this was a bigger piece of news than the advent of the Atmospheric Cabinets themselves. Throughout the 1920s, there had been increasingly loud calls for a refrigerant that was safer than sulfur dioxide, widely used but poisonous, and methyl chloride, equally poisonous and explosive as well. Research chemist Thomas Midgley synthesized Freon in 1928 and demonstrated its safety before a meeting of the American Chemical Society in suicidally flamboyant style when he inhaled a lungful of the gas and used it to blow out a candle flame. Freon was soon proclaimed "one of the most outstanding scientific achievements of our times." That opinion would change, and drastically, but it would take a while.) The new version of the Atmospheric Cabinet made its public bow in early 1932, which was horrendous timing, a period when very few people had money to spend on comfort cooling. And it was a lot of money. The smallest model Carrier offered was priced at nearly $900; installation could run another $500. Almost no one bought them, and the company was forced to abruptly shut down production after it lost more than a million dollars.

The problem wasn't that the public lacked interest; in fact, the magazine *Aerologist* created a minor sensation in August 1931 with the article "Wanted: An Air-Conditioning Flivver!" The "Flivver" was the country's nickname for the Ford Model T, the mass-produced and inexpensive car that had singlehandedly pulled automobiles out of the luxury class and made them available to nearly everyone. And in that vein, the *Aerologist* was hoping for

> an air-conditioning unit for the home, efficient, moderately priced and relatively fool-proof. . . . Its production on a quantity basis by modern manufacturing methods would soon make air-conditioning more of a necessity than the radio or even the automobile, and its acceptance in the home would soon force its general acceptance on a grander scale in practically every other building and conveyance used by man.

But it acknowledged that Willis Carrier wasn't Henry Ford, and that nothing like an "air-conditioning flivver" had yet been developed.

But even with Depression economics, all this attention on the residential market spurred interest in home cooling. Some of the most avid interest came from hay fever sufferers, who were learning that the "cold air" cure so highly touted in the early 1900s owed less to low temperatures than it did to the filtered air provided by most cooling systems. With the spread of air-conditioned movie theaters, doctors often prescribed a double feature for allergy sufferers. And the introduction of home units meant that patients who were desperate enough for relief, and moneyed enough to afford the cost, could avoid the usual therapies—traveling to "hay fever resorts," or going on long ocean voyages—by creating their own pollen-free zones. "Hay fever victims will welcome air conditioning with open arms," wrote *Ice and Refrigeration* in 1934.

However, home cooling had to become more user-friendly. Less than a year after the Atmospheric Cabinet debacle, a number of manufacturers had introduced "console" air conditioners that required no piping, were encased in radio-like veneered cabinets that could be wheeled from room to room, and were almost always depicted in ads that included a model posed with an arm extended sensuously to feel the airflow. Even more innovatively, the first window units appeared. They created their own problem, as the idea of "blocking a window to get fresh air" was still alien enough to most people that *Hardware Age* had to explain it: "They are installed on the window sill in such a manner that the window remains closed and all air entering the room must pass through this device. By this process, the air is filtered and circulated within the room while outside noises are eliminated." And at the highest end of the price spectrum, in 1933 the Chicago World's Fair exhibited air conditioning from a dozen manufacturers, much of it "whole-house" cooling. American Radiator showed air conditioning "for all types of buildings"; the Crane Company demonstrated systems "as applied to the typical average home." And along with the air-conditioned Frigidaire House and Masonite House, there was the glass-walled, twelve-sided House of Tomorrow. Along with its "strikingly modern" decoration—and private airplane hangar—the *Official Guide Book* wrote, "[The living] part of the house is all windows, but none of it opens. The air inside is all conditioned, purified and circulated by ducts." (But the system was allegedly so inefficient that the house's master bedroom was always too hot and wound up being closed to the public.) *The Rotarian* wrote, "The time is now at hand when the better classes of homes, at least, may have complete air conditioning, cooling, and dehumidification for summer comfort and enjoyment."

As complete comfort and enjoyment were a costly proposition for any class of homes, a number of inventors tried to lower the price by dusting off the nineteenth-century idea of providing that comfort to a single piece of the room's furniture. "Air conditioned beds" began showing up throughout the 1930s; unfortunately, they showed up most often in the pages of *Popular Science* and other such publications that specialized in

Illustrations of console air conditioners tended to gloss over the fact that nearly all of them had to be vented to the outdoors, and some required their own water supply line. An ad for the Chrysler Airtemp console encouragingly noted, "Easily installed in a few hours." (New York Public Library, Mid-Manhattan Picture Collection)

not-quite-completely-thought-through ideas. There was a proposal for a coffin-like arrangement of a mattress "surrounded by a canvas wall, but open at the top to allow air circulation," with a tiny air conditioner hovering above the sleeper and blowing down cool air; another gentleman suggested cooling the mattress itself with an "adjustable thermostatic device" and a "motor-driven blower." *Financial World* announced Frigidaire's version, in which "cooled and dehumidified air is served through a porous sleeping bag"; this item seemingly was never manufactured. The Crosley Radio Corporation, which had a brisk sideline in non-radio appliances, came up with the nearest thing to a success with the 1934 Coolrest, "an air-conditioned tent that you can erect over your bed, in your own bed-room!" Even though the Coolrest was priced at a more reasonable $139.50, a fan was still cheaper, and people were very much put off by the notion of sleeping under an opaque covering. A few years later, the tent was changed to a transparent drape, and the name to Koolrest;

The "House of Tomorrow" at World's Fair.

More than 750,000 visitors to the 1933 World's Fair paid an extra dime to tour the House of Tomorrow, depicted in this *Modern Mechanix* layout. As reports had it, the solar heating worked—after all, the house had glass walls—with impressive efficiency. The air conditioning, not so well.
(*Modern Mechanix*)

neither alteration helped much. Within a few years, sales petered out. Only the most desperate summertime sleepers seemed to be willing to bring the Koolrest, or other such alien machinery, into their bedrooms.

With the price of air conditioning firmly beyond the reach of most householders, manufacturers found themselves in the odd position of having to appeal to those very people who had supposedly pooh-poohed conditioned air—the rich. This might be a tough sell. Only a few years before, Mrs. E. F. Hutton had commissioned the ultimate apartment showplace, built to her precise specifications on the top three floors of a new fourteen-story building, that set a record for size. Not only did its fifty-four rooms contain no cooling machinery, but its only concession to summertime discomfort was a screened-in sleeping porch, considerably set back from the noise of Fifth Avenue, for Mr. Hutton's use on hot nights.

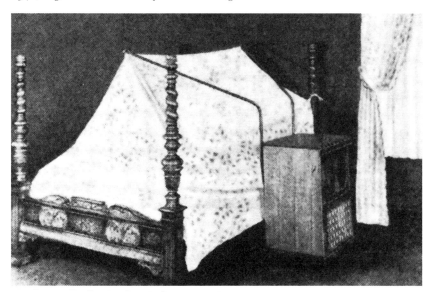

Coolrest advertising tried a variety of approaches. A brochure noted, "Uses ¼ H.P. motor costing only a few cents per night to operate, which is only a small fraction of the cost of electric current for cooling an entire room. . . . It is like going from the steaming jungle to the refreshing north woods." And if it wasn't inviting enough to show the device suspended above a homey four-poster, another photo featured a pajama-clad blonde model preparing to climb in for the night. (Author's collection)

Nevertheless, air-conditioning ads went through a period of reveling in undisguised snobbery. "In homes in Lake Forest at Chicago; Grosse Point at Detroit; Long Island in New York; Greenwich in Connecticut . . . [a]sk the owners of these homes. They'll tell you that since their Doherty-Brehm System was installed, colds and respiratory diseases are infrequent, period furniture doesn't crack, floors and paneling no longer warp, rare editions and paintings are preserved . . . as low as $875, fully installed." *The New Yorker*, which at the time catered to the Park Avenue crowd and had never bothered to advertise anything in its pages more utilitarian than imported scents and large gemstones, now allowed its readers to know that Westinghouse units produced "A Breath of Bar Harbor . . . the sea-cooled, crystal-clean breezes of Maine's smart summer resort. . . . Westinghouse Air Conditioners are completely at home in the most correctly appointed office or the smartest shop. No matter where installed—in appearance and performance, they 'belong.'" And Carrier launched its own assault on the home market by running ads that showed elegant homeowners in full evening dress, admiringly grouped around the air conditioner: naturally, with hands extended to the air flow outlet. The sledgehammer message of the accompanying slogan: "Inside Story: How Leading Citizens Get Relief."

In counterpart to such rich-bitch promotion, there was a sudden proliferation of cheapie units that didn't condition air at all or even cool it but used the name "air conditioning"—and this produced complaints from air conditioning manufacturers. *Business Week* griped in 1934, "Almost anything which has anything to do with the condition of the air is being advertised as 'air conditioning.' One big manufacturer cites as the height of this misbranding the advertisement of a 6-inch window fan selling for $3.95 as 'air conditioning for the kitchen.'" And *The Rotarian*, champion of whole-house cooling, was running an ad for "The Amazing New Scotch Webbing Evaporator" as an "Air Conditioner for Home and Office," priced at only $3.65.

The issue was further confused by machines that cooled the air but had nothing to do with air conditioning. For eons, the American Southwest had unintentionally copied the Indian *tatty* by hanging wet mats in front of open windows to cool incoming air. Once affordable electric fans (and the power to operate them) made it to the area, a number of people had the simultaneous idea to couple them with the wet mats, thus inventing the evaporative cooler—also dubbed the desert cooler, the Arizona cooler, and its most popularly evocative name, the swamp cooler. This was simply a modern version of the *Thermantidote*, using the fan to force air through a wet filter. The 1930s were a golden period for swamp coolers; they were manufactured by a number of companies, could be found in local stores, or were an elementary DIY project. They could be as simple as a slat-sided crate with a fan inside and a piece of sopping burlap slung over its opening, or as elaborate as a whole-house system mounted on the roof, with an automatic water source and ducts extending to every room. In either case, they were a bargain. However, as with the *tatty* and the *Thermantidote*, they worked only in dry weather. Westerners had to become resigned to the fact that, on a damp day (Arizona had its own month-long "Monsoon season"), they would get little or no relief from the heat. And even though swamp coolers were carefully advertised as "Suitable for Dry Climates," plenty of people in more-humid regions were seduced by the low cost—and disgusted by the results.

Then there were room coolers that were out-and-out souvenirs of the nineteenth century. The ChilAire, the Ilg-Kold, the Northern-Breeze, and the Kool-Kleen appeared—even Carrier joined the pack with the Carrier Portable Room Cooler. Unquestionably low-tech devices these were, every one coupling a fan with . . . ice. But in comparison with mechanical units, they were dirt-cheap. Rich's of Atlanta offered the Modine Ice-Fan in several sizes starting at $49.50 and made the offer even sweeter by offering with every purchase "Enough Ice Tickets to Buy 1,000 Lbs. of Ice!" In truth, half a ton of ice wasn't much when it came to satisfying a room cooler's appetite; the Northern-Breeze could consume 300 pounds per day. And there was still that messy problem of disposing of the melted leftovers. Still, *Harper's Bazaar* took a fancy to the Ice-Fan and, in characteristically ditzy fashion, tried to give it a touch of Café Society glamour:

> [It] looks something like two ash-cans fitted into each other with a chromium dome, within which whirs the fan. Seventy-five pounds of ice go into each tank; you turn on the electricity; and lo! the thing begins to generate air fifteen degrees cooler than the atmosphere of the room, reducing the relative humidity enormously. In the dining-room of the Ritz in Barcelona they set huge ice bergs in the garden windows, so that every bit of breeze will come to you over ice. This is the same idea.

Cooling on the Move

All of this media attention may have created lively interest, but it was still the case that household air conditioning accounted for only a tiny fraction of the industry. The vast majority of householders didn't even dream of installing any cooling equipment more elaborate than an electric fan—a situation that would persist for decades. But when mechanical cooling was introduced to the railroads, the reaction was much different.

Since the 1830s, the rise of steam-driven trains had stunned the world (a first-time rider wrote in 1832 of the "novel and stupendous specimen of human skill" that enabled him to travel at the unprecedented speed of twenty-four miles per hour). Within a decade, railroad travel had become the ultimate in rapid transport.

But lightning-fast as they might have been, trains gave passengers almost nothing in the way of comfort. The first rail coaches were little better than boxes on wheels, with candles for lighting, tiny wood stoves for heat . . . and in the summer, no ventilation other than open windows. Anyone traveling in hot weather learned that an open window was a bad idea; it brought in all the dust kicked up by that speed, as well as smoke, soot, live cinders, and an occasional metal splinter blown back from the engine.

By the 1850s, traveling by rail had gained the reputation of being a sweaty, filthy experience, especially as the introduction of ice-cooled refrigerator cars had meant that beef carcasses could travel more comfort-ably than people. But prospective inventors were trying to fill the gap with hundreds of "ventilation" contraptions designed for railroad travel. With straight-faced relish, *Scientific American* described them; an 1852 article went through seven inventions, ranging from "a refrigerator filled with ice or water, which purifies the air above intended for ventilation, there being between the floor of the car and the refrigerator a false bot-tom" to "a long flexible tube, running the whole length of the train from the fire-box to the locomotive, with branch-pipes let into the top of each car, the commencement of the pipe near the engineer being funnel-shaped, so that air can easily rush in." An intriguing idea, but *Scientific American* found a snag: "There is one objection to this plan which struck us particularly, and for which we do not recollect to have seen any rem-edy: should the engine be pushing the train, instead of drawing it, the apparatus would of course be of no avail." There also were attempts at fans, driven by pulleys attached to the car's axle; mechanical ventilators designed to scoop up air at one end of the car and eject it at the other; and Indian railways went historical, installing the water-soaked *tatty* in carriage windows. As few of these devices did much, and none of them did a thing if a train wasn't moving, they gained little popularity.

In 1855, the *Louisville Daily Journal* had been impressed with an early August demonstration of "Barry's Ventilating and Cooling Apparatus," which had been able to lower the temperature of a railroad car from 84 to 76 degrees by means of a fan-and-pulley arrangement, blowing air over 500 pounds of ice ("required for a run of four hours"). The need for that much ice, combined with the high cost of installing Barry's Apparatus ($100 per car), prevented it from being taken seriously.

Throughout the years, Barry's device was aped by a host of other ice-and-fan arrangements that were presented, examined, and found to be not worth the bother. The Baltimore & Ohio Railroad tried in 1884 its own version, which was abandoned after it produced only "poor circula-tion of air in the car and excessive use of ice." The *Railroad Digest* made a longwinded proposition in 1893:

> In sleepers, drawing rooms, and other first class cars, where passengers pay more, special small ice-coolers may be placed in different parts of the car, connected with the main pipe, for summer use, so air will blow through ice; and extending a suitable pliable pipe from the cooler, pro-vided with a spring faucet, and terminating in an enlarged perforated head, fine, sieve-like, it will spray the current into a fine refreshing breeze,

so people suffering more from heat than others, could avail of this spray, and cool themselves to their satisfaction, the hottest day in the year.

That idea didn't take hold, either. In 1901, *Railway Magazine* reported on the French State Railway's "contrivance for cooling railway carriages in hot weather"—an ice-filled box through which air was supposed to blow. It sank without a ripple.

The problem was that railway cars were large enough to become miserably overheated but small enough that the era's technology wouldn't quite fit on board. And that technology probably wouldn't take kindly to being jostled by a train's movement. Electric fans (for that matter, any electric power at all—generators were hefty machines) didn't make it onto a train until the 1890s. And when it came to cooling, the industry standard, the air washer, was far too clumsy a machine to ride the rails. Willis Carrier himself was asked to give some thought to the matter of railroad cooling in 1907. Within the boundaries of what could actually be done at the time, he tried to use his imagination. But even he couldn't come up with anything more effective than the notion of an "apparatus room" under a station platform which would manufacture cold air to be pumped into railroad cars. He promptly scuttled the idea as unworkable.

As the twentieth century progressed, there were sporadic reports of one railroad or another attempting some kind of mechanical cooling. The Great Indian Peninsula Railway was said to have tried a refrigeration system on one train in 1914, a costly and short-lived experiment that was cut short by the beginning of World War I. The Baltimore & Ohio was back at it in 1925, equipping one car with a small-scale air washer; but it drew so much electricity that it was judged unusable. And the Pullman Company, which for decades had been criticized for the stuffiness of its berths, made its own attempt in 1927 when it installed a cooling system in one of its cars . . . only to rip out the system four months later. Train travel remained, as it had been for nearly a century, an uncomfortable— and very grimy—mode of transportation. At most stations, whiskbroom-wielding porters made themselves available to brush off disembarking passengers.

But now, travelers were no longer willing to accept this as inevitable. Throughout the 1920s, railroads lost half of their ridership to automobiles and the newly organized airlines. To regain business, they needed all the luxurious comfort they could get.

So the Baltimore & Ohio asked Carrier to give it one more try in 1929. This time, the centrifugal compressor made it possible to design a system that could not only work but also fit on board. After months of testing, the result was installed in the spring of 1930 in the elegant *Martha Washington* dining car, running on the Northeast Corridor. A meal in a railroad dining car, hotter than any passenger coach with the heat of its attached kitchen, had always been one of the most hellish parts of a summertime train trip; but now things were different. *Railway Locomotives and Cars* reported that "the diner was the most popular car on the train."

To maximize the impact, Carrier organized some shrewd publicity. That summer a "special" train, with the *Martha Washington* as its centerpiece, traveled to Minneapolis for some showoff time at the annual meeting of the American Society of Heating and Ventilating Engineers. Then it went to Atlantic City for the convention of the American Railway Association. Some 3,000 railway men toured the 96 degree train as it sat in the sun, noting with amazement that the *Martha Washington* remained at 73 degrees.

Within months of the *Martha Washington*'s unveiling, the *Santa Fe Chief* got an air-conditioned diner, and the Missouri, Kansas & Texas Railroad installed a few of its own. Then in 1931, the B&O threw the gauntlet in a big way when it took out a series of ads in the *New York Times* that featured an elegantly dressed lady lounging blissfully in a seat while a nearby thermometer registers 72:

> Imagine a train where the temperature in midsummer, on the very hottest day, is always comfortably cool! Where there is no dust or smoke or cinders! . . . where the air you breathe is pure—better than most people have in their own homes! . . . an air-conditioning system throughout—but NO EXTRA FARE![1]

This was *The Columbian*, the B&O's newest New York–to–Washington train, and its success was so thunderously complete that the B&O had to put a second *Columbian* into service within two months. Everyone talked about the train, even recommending it in journals that normally didn't run travel blurbs. A writer for the normally sober *Public Utilities Fortnightly* excitedly advised his readers,

> More or less by chance I took the Baltimore and Ohio's air-conditioned train, and that made it necessary for me to take the five o'clock air-conditioned train back. I would not have traveled any other way. The meanest of God's creatures is a coatless man in a Pullman car, with cinders in his neck and his shirt crawling on him.

The FIRST completely
Air-Conditioned* TRAIN

*COOL *CLEAN *QUIET

The COLUMBIAN
To PHILADELPHIA
and
NEW YORK

(Baltimore & Ohio Railroad Archives)

Even though air conditioning could cost $5,000 per car (in Depression-era money, at that), the increase in ridership was so astounding that railroads scrambled to install cooling on their trains—eight major manufacturers lined up to offer their own equipment, and those renegade lines that tried to pinch pennies with ice-based systems discovered that machine-made cold was actually *one-fifth* as expensive to operate. Barely four years after *The Columbian* started its run, there were nearly 5,800 air-conditioned railroad cars in operation, and more than 12,000 by the end of the decade.

Railway Age would soon write, "Air-Conditioning Had Fastest Growth on Roads," and it was right. Air-conditioned movie houses did spring up everywhere in America, but primarily in America; however, air-conditioned trains sprang up all over the globe, even making it to Australia by 1937. In the world of air conditioning, its adoption by railroads was the standout success story of the 1930s.

This was in direct contrast to the story of automobile air conditioning, one of the decade's most resounding flops.

Even in the days of horse-drawn vehicles, there had been interest in the idea of comfortable summertime travel. In 1884, English carriage builder William Whiteley tried the old idea of an axle-driven fan, blowing air over a block of ice. It got some mild attention, but no more.

The point was made moot when the carriages became horseless; the first engine-driven cars were completely open to the elements. Ventilation turned into an issue only when "closed cars" were manufactured, with windows that had to be clumsily hauled up and down on leather straps. In 1905, *Country Life in America* admitted that a closed car was protection against rain and dust—"but what about ventilation without which a closed car is unendurable in summer time?" To remedy the problem, the writer gave directions for building a "weather hat" to be inserted into a car's canvas top. This required tin sheeting, a six-ply screen, wood battens to prevent the roof from collapsing under the thing's weight, and a "receptacle" to collect any mud produced from driving in rainy weather.

Not much else happened to get fresh air into cars. Windshields were developed that raised or lowered or folded to admit a breeze. The ventilator began to show up in some cars, "an adjustable slide or register closing or uncovering perforations in the body, doors or dash." Otherwise, automobile ventilation was a custom job. When the Maharajah of Rewah ordered himself a luxurious limousine in 1912 equipped with an "electric ventilator fan fitted in the roof," and when a "diminutive electric fan, especially designed to ventilate the interior of a closed automobile" was introduced in 1914, the stories were remarkable enough to make it into print. But such crazed extravagance was unusual. The average driver made do with "ventilators" that were little more than small swing-open doors designed to be banged into the hood, doors, or even the roof of one's automobile. With a ventilator in its open position, air would be scooped into a moving car. As well as dust, precipitation, and insects.

By the 1920s, auto factories had become air-conditioned, if not their product. For those who were griping about ventilation, the big innovation was "adjustable" windows that could be rolled up and down. If drivers still felt a lack of air, in 1921 *Popular Mechanics* suggested that "four ordinary screen-door hooks, attached to the body and doors of an automobile, afford a method for holding the doors slightly open for ventilation." But this, and other jerry-rigged solutions like it, had two major disadvantages. First, as cars were traveling faster and farther on paved highways, the very notion of holding the doors "slightly open" was more than a little unhealthy. And automobiles shared a basic problem with railroads—if a car wasn't moving, the air wouldn't be, either.

Drivers were also discovering that if a car was moving with lowered windows, it would admit not only more air but also an extra-large dose of pollen. This fact was slammed home to a well-to-do Houston businessman, a lifelong hay fever patient. Caught between the prospects of driving with raised windows and Texas heat, or lowered windows and convulsive sneezing, in 1930 he simply decided to have his Cadillac sedan air conditioned. Strapped to the rear luggage rack, a trunk-shaped Kelvinator refrigeration unit, powered by its own lawnmower-sized gasoline engine, chugged away (loudly), producing cool air that was pumped into the car by a fan. *Electric Refrigeration News* was intrigued enough to run a story, complete with picture—"Refrigerated Automobile Appears"—but the general consensus was that this was custom work of the most extravagant kind. And the story received almost no circulation.

For a while, refrigerating an auto seemed to be the stuff of pure fantasy. At the 1933 Chicago World's Fair, the inventor/futurist Buckminster Fuller's Dymaxion Car—egg-shaped, three-wheeled, and said to be air-conditioned—made an appearance. Unfortunately, it was involved in a crash that killed its driver (at the front gate of the Fair, no less) and was immediately dismissed as an insane idea.* Later in the year, *Popular*

---

*Subconsciously or not, the Dymaxion Car would be resurrected at the 1939 World's Fair. The General Motors Futurama exhibit dramatized the "world of 1960"— and according to GM, those citizens would travel in teardrop-shaped air-conditioned cars.

(*Electric Refrigeration News*, reproduced from the *ASHRAE Journal*, September 1999, page 44)

Science, which continued to be smitten with the subject of air conditioning, announced the "First Air-Conditioned Auto." This was another test car, with the refrigeration machinery stowed beneath the floor of the passenger compartment. "Any closed car, new or old, may have the air-conditioning system installed, according to the New York concern sponsoring the invention, which expects to manufacture it in the near future at a sufficiently moderate cost to permit its use even in low-priced cars." This didn't come to pass, which surprised no one; just as all those air-conditioned-bed ideas had proved, *Popular Science* was becoming famous for announcing inventions that were years before their time, along with others that were unfinanced pie-in-the-sky. However, that article made one sound prophecy: "The makers foresee the car of the future provided with air conditioning as standard equipment."

It was obvious that there was a real demand for automotive climate control. General Motors began working on the idea in 1933; but that year, instead of air conditioning, the public was introduced to . . . No-Draft Ventilation. This was nothing more than the small triangular windows in each corner of the car—loftily dubbed Ventipanes—by which GM prom-ised that "fresh air is drawn in . . . [and] used air is drawn out, entirely without chilling drafts on anyone in the car." And it was promoted as an aerodynamic miracle. A series of ads insisted that No-Draft Ventilation would fend off marital spats by keeping women's coiffures unruffled; help smokers avoid annoying other passengers by pulling every whiff of tobacco out of the car; ban drafts that might "take liberties with the necks of old ladies"; and even prevent cigar ash from accidentally blowing into the eye of one's employer. While GM went so far as to tout No-Draft Ventilation's effectiveness by staging demonstrations in front of roaring airplane propellers, no one was fooled into thinking that Ventipanes were much more than the same old open windows, redux.

All the time, the company was quietly hammering away at the puzzle of automotive air conditioning. Some of its thunder was stolen when air-cooled buses appeared on a few southwest routes in 1935, and in Syria a year later.* And the issue was thoroughly confused in 1938 when Nash Motors came out with its Conditioned Air System—which turned out to be, specifically, a fresh-air ventilator-plus-heater designed for cold-weather comfort ("you can drive in your shirt sleeves through sub-zero blizzards") and having no cooling capabilities whatsoever.

But by 1939, GM was nearly ready with an air-conditioned Cadillac . . . at precisely which point it was trumped by the Packard Motor Car Company, introducing the Weather-Conditioner in its new 1940 models.

A flood of jaunty advertising appeared to welcome the innovation. "Note *Air Conditioning*—a Packard first! Real, mechanically cooled *air*

---

*Planes, too. In the very first passenger airlines of the 1920s, planes traveled slowly enough (90 mph), and at low enough altitudes, that passengers were encour-aged to open windows if they wanted an in-flight breeze. And cabins weren't airtight, nor were they meant to be; aeronautics experts pointed out that plenty of oxygen would be provided by air "entering through cracks and crevices." But by 1930 *Popular Aviation* wrote, "The high speed of airplane travel creates an unpleasant draft when ventilation is provided by a plain open window. Again, with open windows the noise of motors and propellers is objectionable, and *there is always the hazard of passengers leaning out and dropping objects from the cabin*" (emphasis added).

That changed in 1934 when United Air Lines introduced the Boeing 247, a stream-lined plane that originated the culture of the red-eye by offering a twenty-hour coast-to-coast flight. The 247 revolutionized the industry: Along with its "three-mile-a-minute" speed, its passengers would be treated to such new luxuries as a cabin whose sound-proofing allowed conversation "in a voice only slightly louder than normal," seats that were upholstered rather than woven wicker, even a lavatory. And for the first time, conditioned air.

conditioning, an optional extra cost feature that pays big comfort-dividends the whole year . . ." "No mere ventilating device, but real, refrigerated *air conditioning*. (Costs extra, but you'll bless it every mile you drive!)"

It surely did cost extra. Adding a Weather-Conditioner to one's Packard came in at $274: in modern terms, over $4,000, a considerable amount in an economy that was still climbing back slowly from the brink. But as Packard had always positioned itself as one of America's most exclusive cars (its slogan was "Ask the Man Who Owns One," and there were other ads proclaiming it "Socially—America's *First* Motor Car"), the company was betting that its well-heeled customers would be happy to shell out for summertime comfort . . . not to mention the nose-thumbing cachet of the words "AIR CONDITIONED" discreetly emblazoned on the hood in chrome.

A number of automotive journals noted the innovation with approval. *Esquire*, appealing to men who were solid members of the Old Boy Network (or wished they were), produced its own highbrow review: "Speaking of the sun and its vagaries, incidentally, brings us with tongues hanging out to the new Packard. For here, in all its pristine glory, is igloo-refrigerated air-conditioning, as complete, as scientific, as comforting as the atmosphere of Radio City."

However, the igloo-type comfort was counterbalanced by a few glitches. While the *Esquire* writeup mentioned that the cooling machinery was "ensconced in the luggage compartment," the truth was that the compressor took up an extremely generous slice of trunk space. As well, it had exactly one setting: ON. The cold-air outlet, mounted above the rear seat, tended occasionally to dribble water on anyone sitting underneath it. Most surprising, and least Packard-like, was the fact that the system had no controls. To turn off the air conditioning in mid-trip, the driver would have to pull over, cut the motor, open the hood, and disengage a belt.

The Society of Automotive Engineers' *SAE Journal* wrote in 1940, "In these circumstances, it takes courage to pioneer a system of air conditioning for all-year passenger car use, and the author feels that those who are pioneering the Packard system are deserving of much praise." But The Man Who Owns One didn't seem to agree. The Weather-Conditioner fell on its face. GM had even worse luck the following year when it finally got around to offering its air-conditioned Cadillac, using a system that was a near-duplicate of the Packard machinery. American buyers who

This ad introducing the Weather-Conditioner ("Cooled by Mechanical Refrigeration in Summer") was comparatively modest. Another version running in *Fortune* addressed its target audience in much more financially explicit fashion: "INVITATION TO BE SPOILED—No other car in America will make you so discontented with anything else!" (Author's collection)

sprang for air conditioning in their cars came almost exclusively from the nation's hottest states.

And there were few of those buyers, at that. By 1942 (when auto manufacturing was suspended completely for the duration of World War II), there were tens of millions of cars on American roads. Fewer than 4,000 of them were air-conditioned.

Towers of Coolness

In the summer of 1929, nationwide papers ran a syndicated article, "Wonderful Skyscrapers for Future," a whole column's worth of wishful thinking. Along with hundred-story heights, blindingly fast elevators, and a plethora of novel conveniences, it promised artificial weather far better than the real thing: "Ultra-violet rays and air conditioning, handled by super-ventilating systems, which will supplant windows and other methods of so-called natural ventilation, will be used to provide healthful and comfortable conditions in these edifices. . . ." The article may have put forth a blueprint for the future, but it was going to take a while.

Since the 1880s, some developers and a growing number of corporations with something to prove, most of them located in Chicago and New York, had been making a public statement by raising structures that were flashier—and higher—than their neighbors. As buildings climbed to ten or even fifteen stories, reactions varied, ranging from "the most remarkable product of nineteenth century architecture" to "a packing case with windows." And as with department stores, theaters, and every other large building, the first skyscrapers wrestled with the problem of ventilation. Windows were vital, not only for air but also for income; various architectural theories insisted that high-rent office space could exist only within twenty-two to twenty-eight feet of a window. And buildings were designed accordingly. To provide the greatest number of windows, some buildings were shaped like a *T* or an *H* or a *U*, while others were solid structures punctured by a central court that was invariably labeled on blueprints "Light and Air." If such an arrangement happened not to admit enough of the promised air, prospective tenants weren't a bit surprised. They had long been used to this lack in every other commercial building.

The twentieth century dawned, bringing with it a golden age of skyscraper architecture, slim, soaring towers that spectacularly transformed city skylines and thrilled passersby—and were no better ventilated than a "counting-house" of the 1850s. Worse, even. If a building's tower was unshielded from the sun, renters would be at the mercy of continuous glaring sunlight. Tenants tried to make do with awnings; they made an incongruously old-fashioned appearance when dozens of them flapped around a tall tower, but they were absolutely necessary. If people wanted more relief, it was up to them to bring along their own electric fans. On particularly grilling days, businesses simply gave up and sent employees home early.

By the 1920s, the advent of air conditioning, and more and more encounters with it in banks, theaters, and stores, created an appetite for better summer comfort. In hot climates, it became urgent. So it followed that the first air-conditioned "skyscraper" would be built in Texas. The trade journal *Buildings* celebrated it in late 1928 with an article entitled "San Antonio's Latest and Largest Office Building the Milam Building, Twenty-One Stories, Reinforced Concrete Frame, Has Many Novel Features, Including a Very Complete Air-Conditioning Plant."

At the time, the Milam Building didn't create nationwide amazement. Its twenty-one-story height was unremarkable; plenty of other cities had skyscrapers twice as tall. And as far as the air conditioning went, anyone could experience it at the movies. But technical journals and industry insiders were impressed. *No one* had ventured to cool twenty-one stories of office space before. Carrier, who had installed the system, pointed to the Milam Building as a major achievement. Independent studies demonstrated the obvious fact that office workers who functioned in cool surroundings were more productive than those who sweltered in the heat. And the building itself touted its cooling system in an endearingly offbeat way when it ran an ad in the *San Antonio Evening News* entitled "Why I Came to the Milam Building," an essay from a businessman who had leased space in the building:

> At the time I moved in I was a little prejudiced against the air-conditioning system. I don't know why, for I had never taken time to investigate it. But after moving in I prize it above all features of the building. I feel better, and finish the day's work with more energy. . . . Then, too, with the windows closed I can keep the wind out. If there is anything that is irritating to me it is to have my papers, telegrams, etc., distributed on my desk and then, momentarily forgetting a paperweight, have some of them blown out of order. Here I can lay the lightest papers about my desk without fear of having them blown about.[2]

Only months after the Milam Building opened, Carrier told readers in the *Saturday Evening Post*, "Architects who have examined this building

Photographer Wendell MacRae caught this view of a modern New York building employing an ancient remedy for heat. (Wendell MacRae, *Summer*, courtesy Philadelphia Museum of Art)

venture the prophecy that within five years buildings that are not Air Conditioned will be threatened with early obsolescence."

It was an interesting prediction, but it ignored one major problem—air-conditioning a skyscraper wasn't cheap (*vide* the much-publicized installations for both the NBC and CBS networks; in both cases, studios were cooled but the rest of the buildings weren't). Not only was the equipment costly, but the ductwork was quite bulky and took up space. And in an office building, where potential income was measured by the square foot, space taken up by ductwork meant a reduction of profit. As well, the taller a building got, the more intermediate machinery would be needed, which would cost more and would reduce more space, which would reduce more income. The competition to build ever-higher structures culminated in 1929–31, when the 70-story 40 Wall Street, 77-story Chrysler Building, and 103-story Empire State Building each knocked off the other in rapid succession to claim the title of World's Tallest Building. Significantly, none of them offered air conditioning to tenants. (Less than a year after the Chrysler Building was dedicated, though, a Schrafft's Restaurant opened at the street level. It installed its own air conditioning, archly pointing this out in newspaper ads.)

But more ominous for developers was the fact that, by the time Schrafft's was pouring its first cup of coffee, the stock market had crashed and the reverberations were being felt worldwide. Few skyscrapers were built, let alone cooled.

One of the few opened in 1932, when the nation's first savings bank, the Philadelphia Saving Fund Society, moved into a new home. As opposed to its previous headquarters in an 1868 faux-European stone palace, this building was a revelation, thirty-six stories of marble, stainless steel, and glass that proclaimed the new architectural concept, International Style. The building's sleek design was countered by tenant-enticing opulence: luscious woods and marbles, Cartier clocks on each floor, custom-designed Streamline Moderne furnishings, RCA-engineered radio reception in every office, acoustical tiles to guarantee quiet . . . and throughout the building, conditioned air.

To make the design congenial to cooling, the building incorporated some design quirks that would be copied millions of times but were jarring when they were first seen: one wall completely without windows (to corral elevator shafts into one location, along with various bits of machinery); a half-height "service floor," plainly visible midway up the building (to house the primary air conditioning equipment); and atop the

The PSFS Building was astonishingly modern in its design, but when it was first built it resorted to the old-fashioned trick of mounting a thermometer (not visible in photo) in a display window to advertise its air-conditioned comfort to prospective tenants. (Photo: Jack E. Boucher)

building (to hide the system's cooling towers), a neon "PSFS" sign with letters twenty-seven feet high.

While one architectural critic fumed against the building's "illogical design" and "ugliness," Philadelphians took to it, and its rental suites were marketed with the ecstatic slogan "NOTHING MORE MODERN." When it opened, the *Philadelphia Inquirer* whipped up interest with weekly ads during July that played up the building's single most modern attribute: Members of the public were invited to take a handkerchief and visit the building's "sample floor," specifically *at the hottest hour of the day.* "When you arrive, wander around, handkerchief in hand. Notice that you do not use it to mop your brow. Trail it lightly across a window sill or desk and behold how little dust you find. Sniff the bracing atmosphere. . . . When you've done all this, then celebrate, in your own way, the modern office building." To cap the celebration, 1,000 prospective PSFS tenants took part in a survey. Fully 90 percent of them listed the build-ing's air conditioning as its most appealing feature.

The building's architect, William Lescaze, must have liked it, too. Less than two years later he built his own house in New York, as remarkably modern as the PSFS Building, and also equipped with central air condi-tioning—an extraordinary rarity in 1934 Manhattan.

Even though the PSFS Building made its bow at the low-water mark of the Depression (some wags claimed that the red neon "PSFS" hovering above the city stood for "Philadelphia Slowly Facing Starvation"), its suc-cess demonstrated that the "luxury" of air conditioning had worked its way into the mainstream. Also into a big business. Sales of air condition-ing equipment in 1932 totaled $8,000,000; five years later, the total was ten times that amount. The *Washington Post* claimed that air condition-ing was "one of the key industries in lifting the country out of the Depression."

As the economy slowly staggered back to life, new buildings of all sizes rose all over the United States, more and more of them with air conditioning. And this was notable only for the fact that it was no longer big news. New York's newest Metropolitan Life building was completed in 1933, a monolithic 30-story structure that briefly tried to make hay out of its status as "the largest office building in the world with 'manufac-tured weather.'" But business gossip was more interested in the fact that the building had originally been intended to serve as the base of a 100-story skyscraper, a plan that was quite literally cut short in the aftermath of the crash. And when Chicago's Field Building opened a year later, the structure's odd air conditioning plan—of its 45 floors, only the first four were cooled—was of considerably less interest than the fact that the developers had decided to build anything at all.

The stories on air conditioning that made the news tended to have another angle. *Time* reported on the June day in 1933 when Chicago hit a record 100.1 degrees and "Col. Robert Rutherford McCormick, publisher of the *Chicago Tribune*, gave thought to his employees sweltering in that magnificent Gothic pile on Michigan Avenue, the Tribune Tower." Only a day later, the *Tribune* announced that the Tower was going to be air conditioned. Aside from the enormous cost of such a project, it was surprising news for a number of reasons. The building was nearly new, having been completed in 1925; "retrofitting" was unusual enough that the word hadn't even been coined; and from the day of its opening, the Tower had already been dealing with summer heat in the old time-honored way—tenants were provided, free of charge, with awnings. Nevertheless, Col. McCormick was as good as his word. A year later, the job had been completed, so successfully that the Tower's landmark neighbor, the Wrigley Building, decided to install air conditioning in 1936.*

Those two projects were great successes, if only because air conditioning had finally been accepted as a cut-and-dried science: Follow the rules and the system would work. Even at this stage, though, occasional problems resulted when someone tried to cool a building with less-tested methods. A notably dismal failure of this kind came in 1933 from the hand of the superstar architect Le Corbusier (who, at least one account claimed, was unfamiliar with the "progress" of Willis Carrier) when he designed, for the French branch of the Salvation Army, its Paris hostel Cité de Refuge. A notable feature was that the building had been designed with a front wall entirely of windows, to give its residents "the ineffable joy of full sunlight." Another feature was that the windows were designed not to open. To keep the building cool, he proposed *l'air ponctuel*, "regulated air"; the windows would be double-glazed, and refrigerated air would be circulated between the panes to provide a thermal barrier against heat. However, because of budget problems, the double-glazing was omitted. So was the refrigeration. The building opened in bitingly cold December weather and was acclaimed for its cozy warmth. But as

---

*Then there was the *Tribune*'s New York counterpart, the 1929 *Daily News* headquarters, also a newspaper building that had been built without air conditioning, but was going to stay that way. *That* building's ventilation came from windows—as one story went, carefully designed to be narrow enough that a stenographer could raise them without undue strain.

winter turned to spring, it continued to be warm. By the summer, it was suffocatingly hot. The glass wall had to be torn out and replaced with one that included windows which could open. While it meant that the residents could finally breathe, it meant that there would be no chance to test the theory of *l'air ponctuel*.

As for Le Corbusier, he was reportedly upset at seeing the operable windows. They had ruined his design.

This story was an anomaly. By the mid-1930s, nearly every day brought news of buildings that were improving their climate control, ranging from the Corey Hill Hospital in the Boston suburb of Brookline to Walgreens drugstores across the nation. And there was Frank Lloyd Wright's 1939 Johnson Wax Administration Building in Racine, Wisconsin, another "hermetically sealed" brick building, but this time equipped with full air conditioning.

Many of the buildings now being cooled were hotels; travelers had finally realized that if they could spend a trip in an air-conditioned train, they had the right to demand the same comfort in their lodgings. While the St. Regis was easily able to update its groundbreaking 1904 system in 1937 by using the original ductwork, other hotels faced more of an upheaval. For those buildings, Carrier announced the Conduit System, which made use of a narrow-diameter duct that allowed tall buildings to be cooled without the need to make costly alterations. One of the largest hotels to take advantage was the Waldorf-Astoria, which had opened in 1931 with only its public rooms cooled; by 1939, it was able to advertise itself as "The World's Most Extensively Air-Conditioned Hotel."*

In this case, "extensively" was very much a relative term. As with nearly every other hotel in the United States that installed cooling, many, but not all, of its rooms were air-conditioned, which was a frank but awkward cost-saving measure. These hotels compensated for their trouble by offering coolness as an extra-cost option. A listing of Kansas City hotels gave a careful rundown: "Air-conditioning is available, fifty cents and one dollar extra, at the Muehlebach and the Pickwick; no extra charge at the President, and one dollar extra at the Statler."

Not only hotels, but apartment houses were springing up that included cooling, from a 1929 building in Spokane to the Edificio Kavanagh in Buenos Aires, an extravagantly Art Deco skyscraper that was at

---

*One of the more famous spaces that remained non-cooled was the Waldorf's top-floor nightclub, the Starlight Roof. True to the room's name, its ceiling could roll back to expose the night sky—a romantic touch, but frequently uncomfortable on humid

its 1936 opening the tallest building in South America and featured 105 centrally air-conditioned apartments. Of course New York got into the act; one of its luxury apartment houses, 400 Park Avenue, found itself at the wrong end of Depression economics when there were no takers for its twelve-room suites. The building was gutted completely, refashioned into more affordable five- and six-room units—and, in a revolutionary move, centrally air-conditioned. Two years later, another developer went this idea one better when an apartment house was constructed in Manhattan's very tony Carnegie Hill neighborhood. Its two- to four-room units offered a taste of the high life to those on a budget, with such ultramodern amenities as glass block windows, television jacks, and individually controlled central air conditioning to boot.

One observer wrote, "The day is near when a house without complete all-year comfort air conditioning will be considered practically, if not actually, obsolete."

Washington's Hot Air (Part VI)

By the late 1930s, that prediction seemed at times to be coming true, fueled by incessant publicity; Willis Carrier himself had helped to bring his company back into prominence with a series of magazine articles and radio interviews in which he assured the public that "the air-conditioned life" was spreading everywhere.* One could read about air conditioning installed in Long Island's eighty Friendship Homes or Cleveland's two model Kelvin Homes, at New York's swanky 21 Club (which proved a fiasco on its first day of operation because someone had mistakenly set the controls to HEAT rather than to COOL), aboard the *Queen Mary*, in Egypt's Hall of Parliament, or more than 8,000 feet underground in a South African gold mine. Even the phrase itself had entered the language as a sales totem: Women's dress materials, bathing suits, shoes and hats for both sexes, all were advertised as being "air conditioned." And the 1939 World's Fair—"Building the World of Tomorrow"—celebrated the apotheosis of conditioned air, or at least Carrier's take on it, with the Carrier Igloo of Tomorrow. "The only exhibit at the Fair devoted exclusively to the modern magic of air conditioning" provided visitors

----

nights. Or worse. A *New York Daily News* columnist noted that the Roof's suavely dressed revelers were occasional prey to flyover attacks from misbehaving pigeons.

*But, disagreeing with other commentators, *Fortune* was unhappy with the industry's growth. It stated categorically in mid-1938, "Air conditioning remains a prime public disappointment of the 1930's."

with a breathtaking mix of technology and glitz: A gleaming white, seventy-five-foot-tall igloo-shaped structure, it was refreshingly cooled, and pointed this out at the pavilion's entrance with two forty-eight-foot-tall thermometers, one of them showing the outside temperature, the other showing the interior's constant 70 degrees. Inside, visitors could learn about the latest in air conditioning equipment and an amazing array of places in which it was used, "from a factory near the Arctic Circle to a telephone exchange on the Equator." During the Fair's run, 4 million people crowded in to experience the Igloo of Tomorrow.

But while the nation was admiring conditioned air, the nation's capital had been busy politicizing it.

As promised, Herbert Hoover's reconstructed Oval Office had been ready for occupancy in April 1930, equipped with Carrier air conditioning. It was a perfectly logical idea: As Director of Public Buildings and Parks Colonel U. S. Grant III understatedly told the *New York Times*, "The [White House] is uncomfortable, and in the past it was not easy to ventilate properly." However, the system represented $30,000 of taxpayer funds, money that was being spent for presidential comfort only months after the most cataclysmic stock market crash in history.

(New York Public Library, Manuscripts and Archives Division)

The summer of 1930 was even more viciously hot than that of the year before, and it seemed that five minutes after the cooling machinery was turned on the entire United States was reading about it in every conceivable publication. *Forbes* tried to be helpful by implying that the Chief Executive needed protection from the germ-laden *hoi polloi*: "Filtering, washing, and humidifying the air of the President's office quarters will therefore protect not only the most important man in the United States, but will protect his staff from the visitors and the visitors from each other." Other journals went out of their way to work in a reference to the system whenever they could, sometimes with a slightly malicious grin. *Merchandising Week* noted that "while the rest of the city sweltered in 90-deg. heat, President Hoover worked away at his desk in the air-conditioned Executive Office, unmindful of the temperature outside." *Steel* observed, "In his air-conditioned offices the President is formulating unemployment relief for next winter." *The Weekly Washington Merry-Go-Round* pointed out "one addition, an air-cooling system such as every good movie house has for hot weather." Even *Good Housekeeping* entered the discussion when it described "an air-conditioning system in the new White House Executive offices so that Presidents may work hereafter in year-round comfort whatever the whims of Washington weather." *Time*, however, seemed to get the biggest charge out of mentioning the system at every possible opportunity: "President Hoover . . . awaited inside the air-cooled White House office." "The President's air-conditioned office was delightfully cool (70°) compared to the heat outside." "The air-conditioning machine was turned on at the White House. Thirty-five newspaper editors spent two hours talking things over with the President. . . ."

All of that scrutiny gave the unfortunate impression that Hoover was luxuriating in high-priced coolness while the rest of America was sweltering through hard times. It was then compounded by talk that not only Hoover's office but also his living quarters were going to be cooled: very bad timing, coming as it did during the country's financial low point of 1931. The White House instantly squelched any such rumors by informing the press that the Hoover household was going to do nothing of the kind, thank you, and intended to get along just fine with electric fans.

One staunch supporter was disappointed by the news, complaining to the *Washington Post* that "if the President had accepted the plan, it would have not only relieved him of many trips out for rest and fresh air, but would have been an example and source of inspiration for the whole

Herbert Hoover wasn't the only figure affected by the Washington air conditioning fracas. Even Vice President Charles Curtis was enlisted for a summertime publicity photo, uneasily posed in his office with handkerchief and electric fan at the ready. (Photo: Herbert E. French, National Photo Company)

world." Strickland Gillilan, reporting on this in the *Los Angeles Times* in the summer of 1931, couldn't keep a straight face.

> The White House is not to be air conditioned this year. The occupants will have to suffer through the heated term with just the usual refrigera-tion, iced drinks, shady halls . . . Tough! . . . Some day when I get rid of this hangnail and my fever blister, I may try to worry about things like White Houses that are not having their air re-conditioned.[3]

Once Hoover turned over the reins to Franklin D. Roosevelt, things changed—somewhat. There were personal preferences to be considered. FDR himself was subject to chronic sinus ailments, believed that *coolth* would aggravate them, and was no fan of air conditioning. But planning to live with the Roosevelts was FDR's indispensable right-hand man, Louis Howe, who suffered from a battery of illnesses, among them asthma, and was no fan of summertime humidity. The Roosevelt entourage moved into the Executive Mansion in late March 1933. Barely three months later, Washington's summer ratcheted up to full intensity. Westinghouse was called, and six console units, with individual controls, were quickly slapped into the fireplaces of White House bedrooms and offices. FDR's own air conditioner received very little use; the historian William Seale wrote that the top of the unit was "used as a depository of clutter."

Throughout the Roosevelt administration the federal government expanded dramatically, and always into air-conditioned surroundings; Washington had finally absorbed the idea that a comfortable workplace would give an added boost to New Deal vigor. A series of generous allotments, made without complaint by a Congress that had itself been working in cool surroundings since 1929, ensured that all new governmental office buildings would include air conditioning as part of their initial construction. By 1935 the *Wall Street Journal* remarked, "The New Dealers are setting the pace for the entire country. . . . Washington [is] the most highly air conditioned city in America."

Roosevelt's dislike of cooled air, however, complicated any plans to install cooling in his own quarters, in large measure because he demanded the ability to do without it. While government buildings were designed with central systems, the First Family's private space had to make do with a series of "portable" window units (in an ironic coincidence, the White House wiring was discovered at this time to be dangerously inadequate; the Roosevelts had to spend extra time in Hyde Park while it was overhauled). When the West Wing was expanded in 1934 and fitted with an $80,000 air conditioning system, it was divided into "seven separate zones of ventilation." The ventilation in one of those zones, the Oval Office, was rarely used. Even in the most brutal heat, rather than turn on his own air conditioning FDR preferred to open windows and roll up his sleeves. Reporters frequently described the rumpled, perspiring Roosevelt with admiration.

It wasn't only the President's climate-control tastes that made it into the press. Nearly everyone by now had walked into a store whose air conditioning wasn't working, or witnessed a neighbor lady's complaining to the movie-house manager when the theater was too cold; but stories about "air-conditioning politics" turned out to be infinitely more fun when they featured actual politicians. Americans loved reading that Secretary of Labor Frances Perkins had asked for the air conditioning in her office to be kept on until 8:00 P.M. "for her special benefit," that the request had been turned down and she was given an electric fan as a consolation prize; that the air conditioning in one government building had mysteriously stopped working, until it was discovered that the system's cold-water intake had been clogged by Potomac River jellyfish; and the most popular story of all, that the nation's senators continually squabbled about the proper temperature for the Senate Chamber, with younger senators preferring 70 degrees and "middle-aged solons" demanding it warmer. The new picture magazine *Life* even satirized this ongoing argument with the sketch "The Frozen Congress" ("Unnoticed at first, a doorkeeper of the House gallery, after several days, finally peered in the door and discovered the entire membership of the House was frozen stiff. . . .").

One instance of true American pride in governmental air conditioning involved British royalty, when King George VI and Queen Elizabeth visited Canada in 1939 and worked in a side trip to Washington to see the Roosevelts. As this was scheduled for the second week of June, there were worries about a possible heat wave, but press accounts shrugged it off: The White House had recently installed a small-scale central system "to chill a few of the private rooms most in use." The *New York Times* wrote, "In Washington's notorious June heat the thing that the King and Queen will probably enjoy most in the White House will be its efficient air-conditioning plant. . . ." Sure enough, once the Royals arrived (in an air-conditioned train), the nation's capital was broiled by high temperatures. Dauntlessly, they held to their full schedule of events, including a garden party and a State Dinner. While the garden party had been reported as a particularly trying afternoon, especially for men in dress uniforms, the dinner would take place in the White House's air-conditioned dining room.

But there would be a state secret: That night, apparently the air conditioning didn't work. The columnist and Washington insider Helen Lombard would write of the dinner, "No one could foresee that the air cooling system, so confidently installed in the White House not long before, would contribute nothing but nerve-wracking noises to one of the rarest occasions in White House history!"

6   From Home Front to Each Home

At the beginning of 1940, Willis Carrier gave a speech in which he predicted a bright future for custom-built homes, equipped with centrally conditioned air. (He might have felt that a pep talk was needed. A recent survey had discovered that fewer than one in 400 Americans had air conditioning in even a single room.) In the midst of his enthusiasm, there was a moment in which he admitted that the public awareness of air conditioning "has been brought about only by a long, slow process of consumer education." *Slow* was no understatement. From its first sputtering beginnings, comfort cooling had been around for nearly a century; for half of that time, it had been ridiculed as the stuff of science fiction.

But science fiction of late had been demonstrating an intriguing tendency to turn, albeit slowly, into fact. As recently as 1930, *Advertising & Selling* told of "soundproof and windowless buildings that will be supplied with medically prescribed light and weather." It was a concept that produced raised eyebrows—however, within two years Westinghouse had begun research to investigate that exact idea. Then in 1934, the *New York Times* visited the Chicago World's Fair, which proudly boasted a few of those soundproof and windowless buildings, to cover an event at which a group of experts declared that the coming years would be filled with scientific achievements "surpassing the phantasies [sic] of a Jules Verne." Along with such utter impossibilities as a seventy-year human life span, motors powered by solar energy, and worldwide television broadcasts, it predicted "air conditioning of entire cities from central plants." Readers might have chuckled, but by decade's end, it had become obvious that at least some of those predictions were coming to pass. The ultimate showcase for all this wonderment was the 1939 edition of the World's Fair in New York, the "World of Tomorrow," fulfilling a host of prophecies while astonishing visitors with such breakthroughs as the "spectacular method of decorating with light," fluorescent tubing; GE's "Sun Motor," which could actually convert daylight into electricity; even RCA's long-awaited launch of its own television station. And when it came to buildings of every kind, with "prescribed weather," this Fair's exhibits were using *seven times* the cooling that the Chicago Fair had installed. Now there were predictions of "fabricated" homes, ready to set up in days, complete with centrally manufactured weather machinery built right in. Maybe it was indeed only a matter of time. . . .

In 1940, however, many people stopped happily looking forward to what the future might bring.

War was already underway in Europe, nations that hadn't become involved were bracing themselves, and ventilation was being used in ways that its inventors probably hadn't imagined. It was global news that Vatican City was in the process of constructing air-conditioned bomb-shelter space in the bowels of a 500-year-old tower; at the same time, it was more of an open secret that Adolf Hitler had superintended the construction of his own massive *Führerbunker* on the grounds of the Reich Chancellery. Hitler was an asthmatic, and his demand for not only effective but gas-filtering air conditioning wasn't to be trifled with; at another of his residences, he was infuriated one night by the temperature and had the ventilating engineer brought in (by Gestapo agents) to make after-midnight repairs. Less generally known, Hitler's friend Benito Mussolini, supposedly in a spirit of competition, had authorized the construction of *twelve* bunkers for his own use beneath Roman buildings. These were air-conditioned as well, if less luxurious; one bunker used two stationary bicycles to power the generator.

On the Allied side, things were kept considerably more quiet. It was top-secret information that the British government had fitted out, in the basement of a building conveniently near the Prime Minister's office, its own Cabinet War Rooms, enabling operations to continue even through the Blitz. And although conditioned air was an extreme rarity in London of that era—at least above ground—the complex was fitted out by Frigidaire. *Refrigeration World and Air-Conditioning Review* would report some years later that the system could "cool and dehumidify 80,000 cu. ft. of air per minute throughout the six-acre steel and concrete fortress with its mile of corridors and 500 rooms"; however, perhaps Frigidaire hadn't counted on the amount of coal dust in London's atmosphere, or Winston Churchill's ten-cigars-per-day smoking habit. Valiantly, and not too successfully, the system struggled to do its job. Secretaries tried to filter the incoming air by covering the vents with curtain netting, which would turn black in three days; Frigidaire had to bring in additional Room Coolers to supplement the system; and the War Cabinet Room, scene of lengthy meetings and incessant tobacco usage, was equipped not only with multiple air conditioning vents of its own but with an additional set of electric fans mounted on the walls, presumably to chase along the smoke.

(© Imperial War Museum [*detail*])

In Washington, it was also kept quiet that the Roosevelt administration had a couple of shelters, too, both equipped with their own gas-filtering ventilation: one, a former vault beneath the Treasury Building that had previously been used to hold impounded opium, the other a custom-built structure under the new East Wing of the White House and intended for presidential use (and that FDR reputedly saw only once, after which he announced that he'd prefer to take his chances on the surface).

At the time, the United States was officially neutral in the war, so Americans were concentrating on a "quest for adequate industrial pre-paredness." Previously unemployed people found themselves snapped up for manufacturing jobs, and the fabrication of houses had been set aside in favor of the fabrication of warships. By the end of November 1941, the U.S. Maritime Commission announced that the whole program was proceeding beautifully.

But a week later, Pearl Harbor was attacked. America officially entered the war, and military production clicked into high gear.

Every business, every industry retooled to help the war effort. Automotive firms began producing military trucks and jeeps; Ford made aircraft engines. IBM switched from typewriters to carbines. C. G. Conn stopped supplying schoolchildren with musical instruments and began making altimeters. Even Coty diverted production of its usual makeup shades to concentrate on camouflage paint for jungle fighters.

Air conditioning went to war, too. In 1917, mechanically tempered air hadn't been used much in wartime production apart from its importance in munitions factories. Now it was vital. Many plants operated round-the-clock in blackout conditions, with no open windows, and had to be ventilated. Some machining operations required temperature control to produce metal parts to precise tolerances. Lower temperatures were needed to make a host of other items, ranging from synthetic rubber to high-octane gasoline. Ships needed refrigerated storage; aircraft intended for high-temperature locales needed to keep crew members comfortable enough to function; and submarine air, which in the past had been so unbreathable that subs had been nicknamed "pig boats," needed to be vastly improved.

To meet these needs, manufacturers concentrated on equipment for military use. Carrier Engineering produced "special portable air conditioners" for aircraft and developed a wind tunnel that could simulate high-altitude temperatures as a training aid, while York supplied refrigeration for Liberty ships. Civilian production of air conditioning equipment, though, was virtually halted in 1942 when the War Production Board instituted Order L-38. "L" stood for "Limitation": The order prohibited the installation or manufacture of new air conditioning units or equipment solely for personal comfort—which the *Wall Street Journal* defined as "units for motion picture theatres, public auditoriums, hotels, department stores, office buildings and other unessential uses."

Only months later, in a meeting that was extensively covered by the press, the Board very politely asked the heads of air-conditioned department stores if they wouldn't mind lending their own compressors to war production. Even though it was sentencing those stores to some very uncomfortable summers, Macy's and Gimbels, along with Sears, Roebuck; Marshall Field's; Tiffany's; and a clutch of other retailers—including the first air-conditioned store, Hudson's—immediately volunteered their equipment to less-glamorous but vital locations such as Pratt & Whitney in Kansas City and B. F. Goodrich in Texas. There were even a few rumbles that the members of Congress themselves might be asked to donate the House and Senate compressors to the war effort; somehow, nothing came of it. The War Production Board itself had nothing to offer; its offices had never been fitted out with air conditioning.

For the next three years, the private citizen who owned an air conditioner held his breath every time he turned it on. Not only were repairs and replacement parts extraordinarily hard to get, but Freon—by this time, the refrigerant of choice—had been diverted to the armed forces.* While regulations made Freon available for an ailing family refrigerator, it was nearly impossible to get a recharge for a tired air conditioner. In response to the shortage, electric fans became ubiquitous. And ice made an uneasy comeback. The City Ice and Fuel Company's "Victory" icebox was so popular that the company couldn't keep up with the demand. As well, there were movie houses and legitimate theaters across the country, confronted with broken-down air conditioning, that had to revert to ice-and-fan systems. The market became so frantic that there was a nationwide ice shortage in the torrid summer of 1944. The heat as well as the shortage were both so severe that even Broadway's mega-hit *Oklahoma!* suddenly had seats available, wrote the *New York Times*, "if you cared to see a show in a steam bath."

But at the same time, the end of the war was in sight. The *Wall Street Journal* ran a front-page headline in August: "Cooler Tomorrow/Air-Conditioned Homes, Autos, and Offices Goal of Equipment Makers/They Plan for Post-War Sales 500% Above Their Best Peacetime Year":

> Men who make air conditioning equipment . . . plan to do something cool for everybody except the man in the street. The man in bed will enjoy refrigerated sleep. The man bound for work may ride, within a few years, in an air-conditioned trolley or subway. At his office or factory he will be synthetically chilled. And the motorist may soon be able to have a cooler in his auto.[1]

This wasn't just a case of misplaced enthusiasm. After a Depression's worth of housing insecurity and a war's worth of severe housing shortage, there was a nationwide hunger to own a house; *The American Home*

---

*Nonflammable, and completely nontoxic (so everyone thought), Freon turned out to be the ideal propellant for insecticide "bombs" that were sent to the tropics. The military had not only discovered a new use for the gas, but in the process invented the modern aerosol can.

A notable exception to the shortage of *coolth* came during the summer of
1942, when Carnegie Hall not only tried a season of staged operettas but
installed (temporary) cooling as an added attraction. The *New York Times*
was intrigued at the possibility of behind-the-scenes negotiations to get
the machinery: "What with priorities, the Carnegie Hall fathers had to
exchange some valuable equipment of their own in order to obtain an
air-conditioning system." (Carnegie Hall Archives)

surveyed its readers and discovered that 32 percent of them intended to buy a new home within two years of the war's end. And the sci-fi prophecies of the 1930s had resurfaced, with a sudden emphasis on the "dream homes of the future" that would be made possible by wartime research. The headlines alone intoxicated readers: "A Revolution in Post-War Living Is Predicted," "War Seen Creating New Marvels for Our Homes After the War," "Wealth of Novelties for Civilians," "Wonders at Hand for Daily Living." With the help of "stainless paints," miracle fabrics, and plastics such as Plexiglas (along with the ones that would fall by the wayside: Fosterite, Cerex, Beautanol, Duranol, Pyralon, Lumite, Velon), new homes were going to boast such marvels as built-in television receivers, "electronic tubes which can broil a steak in three seconds"—and, of course, conditioned air.*

There was so much hyperbole that the National Association of Home Builders began to wonder what potential customers were really expecting. Some 500 families who might be in the market for a postwar home were surveyed as to their requirements. Alarmingly, 72 percent of them expected to find an *affordable* new house, "within ten months after the war," built with central air conditioning. By July 1945, *Life*—which was turning into a weekly chronicle of mid-American faddism—gave this fantasy its own stamp of approval by running a four-page photo spread: "AIR CONDITIONING/After the war it will be cheap enough to put in private homes": "One of the prewar luxuries which seems most likely to come out of the luxury class and into the postwar mass market is air conditioning. Manufacturers are now working on home cooling units which, produced in quantity, can sell at moderate cost."

A month later the first atomic bombs were exploded, Japan surrendered, and the war was over. As the general celebration quieted down, wartime restrictions were lifted, things got back to normal, department stores got their compressors back, and the *New York Times* gave a hint

---

*A couple of those ideas would be much too futuristically combined in late 1945 when Plexiglas toured nationwide department stores with its "Dream Suite" exhibition, a plush bathroom-with-tiny-bedroom that used Plexiglas in everything from a shower enclosure to towel bars, seating, wall coverings, and a hat rack. But Mary Roche, home editor for the *New York Times*, was particularly struck by one use: "The proposal: an alcove-like corner of the bedroom just big enough to house a bed, night table and chair and completely closed off by a transparent wall of curved Plexiglas. The point: a sleeping cell small enough to allow for complete air-conditioning at a small cost.

"Plexiglas, in case you've forgotten, is a clear transparent plastic. . . . Up to now it has been used to make bomber noses."

that air conditioning might be something less than a guaranteed instant success when it ran an article, "The Shape of Things and Goods to Come." The full-page spread was adorned with a cartoon depicting all sorts of wartime-restricted goods that were soon to return to the market, ranging from nylons to automobiles. One of the goodies was an air conditioner bearing a sign: "Just Name Your Climate!" Significantly, in three pages of text, the article didn't mention another word about it.

Comfort in Your Own Home (Once More)

If conditioned air was going to be a part of postwar housing, it would have to make the difficult shift from gilt-edged fantasy to homespun reality, and it would have to do so quickly. Starting in 1946, the nation experienced a genuine housing boom, the first since 1925. Most of the customers were returning servicemen, G.I. housing loans in hand and anxious to put down roots.

At first, there were some desperation-tinged attempts to produce the fabricated housing that everyone had been talking about. One supplier went so far as to advertise war surplus Quonset huts, $398 each, as "The Ideal Answer to Your Housing Problem"; indeed, 750 of them were turned into a temporary housing project in Los Angeles.

If that concept was too spartan, prospective homeowners could look to Buckminster Fuller, inventor of the ill-fated Dymaxion Car. Since the 1920s, Fuller had been working on an idea for a lightweight, dome-shaped house, with an aluminum shell and Plexiglas windows, that could be erected almost anywhere. By 1946 it was ready for production, rechristened the Fuller House, and unusual enough that it snagged an article in *Life*: "Newest answer to housing shortage is round, shiny, hangs on a mast and is made in an airplane factory." For $6,500, buyers would get a two-bedroom "machine for living," equipped with a "passive" ventilation system that was touted to change the air every six minutes. This system, an exact copy of the "thermo-ventilation" setups of the nineteenth century, was designed to pull cool air up from ground level into the "house" and eject hot air through a central vent. It happened to be necessary— none of the windows could be opened.

"Builders hope that [the] convenience and comfort of [the] Fuller house will help overcome prejudice against appearance," wrote *Life*, and the writer had a point; the Fuller House looked exactly like a flying saucer (a term that hadn't even been coined yet—at the time, the House was described as a "giant aluminum hamburger"). Although many people were put off by its strangeness, the company reported some 3,500 orders.

MODEL OF HOUSE WITH VENTILATOR

The Fuller house, highlighting its "Ventilator." It was designed to rotate much like a weathervane in order to take advantage of favorable breezes. (Courtesy: The Estate of R. Buckminster Fuller)

They were never filled. Fuller himself, afraid that the design hadn't been perfected, pulled the plug on the venture.

At the other end of the taste spectrum, the California-based magazine *Arts and Architecture* announced its Case Study House Program, an attempt to introduce high-end design into affordable housing. The program started in 1945 with eight "nationally known architects, chosen not only for their obvious talents, but for their ability to evaluate realistically housing in terms of need," commissioned to "create 'good' living conditions [and] contemporary dwelling units." The result was a series of houses, nearly all of them built around Los Angeles, that exemplified modernist architecture: flat-roofed, single-story structures, every one featuring entire walls of glass. Summer comfort was furnished by white, sun-reflecting roof surfaces; by generous eaves that created deep overhangs to shade the expanses of glass; and by plenty of cross-ventilation. The *Los Angeles Times* had mentioned that one house would be air-conditioned. Once the house was built, that apparently proved to be unnecessary.

Although Case Study Houses were supposed to be a solution to the housing problem, they weren't. From a comfort standpoint, their advantages wouldn't easily be transferable outside of southern California's airiness. In the wrong region or even facing in the wrong direction, a flat roof and glass walls could absorb a frightening amount of heat from the sun, wreaking havoc with temperature control and making any one of these houses a very uncomfortable place in which to live. Exactly *how* uncomfortable was inadvertently demonstrated by the "Less Is More" architectural autocrat Mies van der Rohe, when—at precisely the time the Case Study project was taking off—an Illinois client asked him to design a weekend retreat. He produced the Farnsworth House, a visually arresting structure whose roof was flat, whose every wall was glass, and that seemed to float above its rural setting. Architectural critics were bowled over. Art lovers made pilgrimages to see it. Unfortunately, however, the day-to-day experience of actually living in the Farnsworth House presented a problem. Although it stood "in the cooling shadow" of a maple tree, daytime sunlight baked the structure unmercifully, and the owner discovered that the (single) door couldn't be opened for air in the evening because the smallest light bulb instantly transformed the house into "one huge mosquito and moth lantern." The word "unlivable" appeared, and in print, to describe the place. The owner became disgusted and wound up installing screens for insect control. Mies in turn became disgusted with the owner, to some extent because of the screens. The whole argument ended in recriminations, lawsuits, and a surprising amount of negative publicity. To cap off the tale, in the midst of Cold War hysteria a few years later the Farnsworth House's design would be accused of having un-American influences.

When it came to the prospect of plunking down cash to buy a modernist house, a Fuller House, or any other such innovation, it was soon evident that the average home buyer was willing to visit but didn't want to live there. By 1948, *Popular Mechanics* insisted that "the ex-serviceman . . . wants a conventional house as he knows it. He is not satisfied with a converted streetcar, a concrete-clad igloo or a hut-type dwelling even if it is made of platinum. He wants the kind of a home that he has dreamed about at odd moments for years, a place to be proud of where he can raise his family and enjoy life."

Contractors quickly caught on, realizing that production-line ideas, coupled with strictly conventional design, would be the easiest (and most profitable) way to accomplish this. Almost overnight, Panorama City sprang up on the West Coast and Levittown on the East, transforming gigantic parcels of ex-farmland into instant cities: thousands of houses, absolutely affordable, nearly identical, and extraordinarily cost-cutting in their construction. Levittown was particularly noteworthy, as its builder, Levitt & Sons, had originally made a name during the 1930s erecting

Mies van der Rohe's Farnsworth House. At the far wall opposite the door were two small windows. Apparently they didn't help much when it came to ventilation. (Photographs in the Carol M. Highsmith Archive, Library of Congress, Prints and Photographs Division)

distinctive high-end homes, many of them featuring central cooling. But Levittown was an entirely different proposition. Built on concrete slabs, its tiny two-bedroom homes could be erected in mere hours. Even so, they offered a clever touch of extravagance, featuring luxuries that buyers might have only dreamed about before the war: a fireplace, a Bendix Automatic Washer, and, after 1950, the ultimate indulgence—a built-in Admiral television set.

But as to air conditioning, no.

Levittown's sales center was air-conditioned; its 17,447 houses weren't. As contractors and homeowners were discovering, the prepackaged, streamlined, *inexpensive* central air conditioning that had been so confidently predicted as standard equipment in new houses . . . simply didn't exist. There was general agreement that whole-house cooling added roughly $2,000 to the cost of a single-story home. That was a sizeable chunk of income to throw on top of the $7,900 price tag of a Levittown Cape Cod, or even the $9,000 cost of a Panorama City ranch.

Better Homes and Gardens, supreme authority of American middle-class home trends, assured its readers in early 1949 that "Your New Home Can Be Designed for Air Conditioning":

> It's no longer a dream, but something you should consider when remodeling or building, just as you do indoor plumbing and central heating.

But there was a caveat:

> If you're remodeling your house fairly extensively, air conditioning with a central system can be installed with reasonable economy. If you don't plan a major remodeling, it may be impossible to install an efficient system. Portable or room air conditioners may then be the answer.[2]

Tellingly, there were no air conditioning ads elsewhere in the issue—but there was a picture essay that showed the man of the house how to fix an electric fan.

Later that year, after a fierce summer, *House Beautiful* got into the fray with a blunt article entitled "So You Think You're Comfortable!" Had new homeowners felt a great deal of hot-weather discomfort in their houses? No surprise. It placed the blame not on air conditioning or the lack of it but on heedless design, planning, and building: "We have built, in the South, little, low-ceilinged hotboxes without properly shaded roofs. We have built, in the North, houses with thin brick walls that are cold in winter and hot in summer. We have put Cape Cod houses into climates that are as different from Cape Cod as Cape Cod is from England." It was absolutely true. And it was advice that would be roundly ignored, especially by owners of brand-new Cape Cod houses in Levittown and imitation Levittowns everywhere.

At the beginning of the 1950s a few developers began, cautiously, to include central air conditioning in some of their houses. They were delighted to find that it was seen not as useless frippery but as a desirable feature. One magazine article told of a (possibly hypothetical) "three-bedroom ranch" that was put on the market, undistinguished except for its air conditioning; it was snapped up in two hours. Central cooling was proving to be enough of a draw that the *New York Times* wrote in 1953, "Builders are swinging toward air conditioning because of changing attitudes. . . . Some banks and other lending organizations recognize officially now that air conditioning helps maintain the resale value of a home. They are taking this into consideration when granting mortgages." Manufacturers took advantage of this with ads that bore headlines such as "Will your new home be *OBSOLETE* . . . even before you move in?"

And the *National Real Estate and Building Journal* reported an even more powerful factor that attached itself to central cooling: "*Prestige—* There is a certain amount of respect paid to owners of air conditioned houses." True enough. When another Levittown sprang up—this time, in Pennsylvania's affluent Bucks County—Levitt contracted with Carrier in 1956 to provide systems to its $18,000 deluxe model. While the original no-frills Levittown house had been named the Cape Codder, this one was christened, aptly, the Country Clubber. In Levittown society, ownership of a Country Clubber was naturally mentioned as part of one's credentials when one was introduced to others. As a longtime resident recalled, it was equivalent to a gent's admitting that he was a Navy commander.

At the same time, though, central air conditioning was shedding its luxury status in an attempt to enter the mainstream of home building. The trade magazine *House & Home* devoted twenty-four pages of its March 1954 issue to a lengthy article that explored the question "What's new in air conditioning?" As well as nuts-and-bolts issues, it pointed out that "women have the most prejudices against it, yet stand to benefit the most" and offered tips on selling it to wives, such as suggesting that cooking and sleeping (as well as even stickier tasks, such as applying makeup and sliding into a girdle) would be easier in air-conditioned surroundings. Meanwhile, the prefabricator-contractor National Homes, famous for its claim that it could erect a house nearly anywhere in the United States in ten days, was targeting buyers more directly with full-page ads in slick magazines: "All This and Air Conditioning Too!" The ads took advantage of the fact that the price of central cooling was starting to drop; National offered it as a $500 option. Other contractors dangled the idea of air conditioning before home buyers but used the system as a "loss-leader," absorbing part of the price by cutting corners: less insulation, less heat-deflecting attic space, bathrooms with no windows, and particularly "sealed" windows, mere panes of glass in a wall with no way to open them.

The fact that these little economies would wind up making a house hotter in the summer and tough to air out in any season, eternally dependent on its air conditioning, and absolutely unlivable without it—for instance, if the power went off—wasn't discussed. Provided the windows were picture windows and preferably numbered more than one, most buyers were thrilled with the tradeoff. ("Wide, wide expanses of glass—

for that wall to wall treatment so greatly admired!" gushed one window company.) *House & Home*'s article had showcased a suburban home built with picture windows in every room: "Fixed double glazing throughout the house seals it against constant menace of blowing dust. . . . [The only] openings to outside are front door, kitchen and living-room doors. . . ."

Architectural engineers called these virtual walls of glass "heat entrapping." Nearly everyone else called them The Latest Thing.

Just Like the People Next Door

Possibly the first courtroom case involving noise from a home air conditioner began in the summer of 1953, when Brooklynite Sam Arkow installed a "room cooler" in his bedroom window. Its noise disturbed his neighbor across the air shaft, Esther Gershberg. She complained; he moved it to another window. She still complained; he balked. She called it "detrimental to health and a public nuisance." And sued.

The thoroughly amused *New York Times* and *Daily News* got ringside seats as the case zigzagged through the Brooklyn court system throughout July, beginning with "the superheated atmosphere of Flatbush Court, which distinctly is not air conditioned" (and whose magistrate sighed that *he* was having trouble sleeping because of the noise from his own neighbor's air conditioner) and ending nearly a month later when the case was dismissed with a prediction that "the builders of air conditioners will eventually, inevitably make them absolutely noiseless." During the case, Mr. Arkow had been represented by a heavyweight law firm "understood to represent big air-conditioning interests," but those lawyers discovered that Mrs. Gershberg was no pushover. Even after an air conditioning executive had come to her home, listened to the offending unit, and offered her an air conditioner of her own "to drown out the noise," she refused categorically. As she announced to the court, "I wouldn't let him bribe me that way."

Considering that the summer would end in a heat wave that broke records over half of the United States and extended all the way to India, with ten consecutive days of 90-plus–degree temperatures in New York alone, perhaps she should have taken the air conditioner. And there was also the fact that Mrs. Gershberg's "public-nuisance" claim was quickly losing ground.

The industry had never swayed from its position that central air conditioning was the only *real* air conditioning. And "portable" units were still derided by engineers as a Band-Aid–type remedy, unable to accurately control humidity or to provide absolute temperature control. The public couldn't have cared less. Such ideals were fine for an office build-ing—but when it came to their own homes, and the prospect of getting some uninterrupted sleep on a hot summer night, perfectly controlled humidity was less than crucial. Moreover, if those homes were rented houses or apartments, central air conditioning, and particularly its duct-work, was a ridiculous suggestion. The room air conditioner was their only alternative.

After the war, production of air conditioners had resumed, and right along with it the old complaint that they were too expensive. Throughout the late 1940s, the average window unit hovered near the $400 mark (in modern terms, nearly $3,500), was usually advertised at a $50 discount, and didn't seem to be able to drop any further in cost. Even the *Times*'s Mary Roche, whose Home Section articles were designed to blithely push household goods no matter how exorbitant, acknowledged the steep price in 1949 when she wrote "Cooling Off with Electric Fans" for those "who cannot afford the $350 it takes to buy a window-sill air conditioner." While the article served mainly as puffery of the newest "styles" of fans and tips on the best ways to use them, its last paragraph admitted that electric fans couldn't lower temperatures (readers could easily agree, as that summer had been brutally hot). And it offered a rather extreme solu-tion: "But if the atmosphere starts to sizzle the day twenty guests have been invited for cocktails and supper, more drastic measures may be nec-essary. One is to buy a hunk of dry ice, set it on the floor in an out-of-the-way corner, or on a low table, and place a fan behind it. A cubic foot of ice will last about three or four hours and do a surprisingly good job of air conditioning."*

Truth was, if air conditioners had been seen as life-changing objects of desire, they now had some real competition—television, which was exploding into living rooms everywhere and literally transforming daily life. Outside activities, mealtimes, furniture arrangements, even furniture *designs* were changed to accommodate the new "electronic hearth." And nearly half a billion dollars in sales proved it. For a few years there was a tug-of-war for the attention and wallets of the public, and it was under-standable; with an air conditioner priced at a minimum of $350 and a

---

*In a room with closed windows, not a good idea. In 1941, *The New Yorker* had told of a banker who thought his conference room could be cooled in exactly that way. But as dry ice is composed of carbon dioxide—which is released as the stuff is exposed to air—the whole group nearly lost consciousness. However: "Before any of the gentlemen keeled over with carbon-dioxide poisoning, the conference was broken up by the appearance of a minor employee who knew something about chemistry."

television set available for less than $200, most people were opting for the TV. At the end of 1950, television sales had rocketed beyond anyone's wildest expectations, while air conditioning representatives glumly reported that "production far outstripped demand."

Nevertheless, the two appliances had been romantically linked in the public imagination. The *National Real Estate and Building Journal* observed, "Television is becoming the family recreation center, replacing the fireplace. The comforts of air conditioning and the entertainment of television are offered as means of bringing members of families closer together." The *New York Times* agreed, noting that "television sets have become the companion item in popularity, supplying in the entertainment line what the air-cooling machine furnishes in the line of comfort for the family." The *Hartford Courant* made the relationship logical by asking, "How are you going to sit home and watch television this summer if your living room gives you a Turkish bath treatment?" As if in response, Fedders promised to deliver "a pleasant afternoon watching baseball on TV when it's blazing hot outside."* And RCA, which also had entered the air conditioning field, went so far as to build an entire ad campaign on the supposed link with television when it urged customers to "'Tune in' perfect weather with your RCA Room Air Conditioner."

There was another angle to this relationship, one that no manufacturer dared to mention; a TV set, with all its tubes, spilled a remarkable amount of heat into the average living room. This made air conditioning even more welcome.

The momentum was slowly building, but fate accelerated things during the summer of 1952 when much of the northern hemisphere was broiled by a series of heat waves that stretched from Mexico to Moscow, and the consumer guide *Kiplinger's Personal Finance* predicted that the air conditioner was "the machine most likely to become your next household necessity. . . . Coolth is the next big home product." Sure enough, before the end of the year, more than 365,000 air conditioners had streamed out of American stores. The following year, when Mrs. Gershberg hauled her neighbor into court, temperatures climbed even more cruelly, staying high into September, and the media started referring to the

---

*This relationship even made it into mainstream fiction with Ray Bradbury's 1951 sci-fi tale "The Pedestrian," a futuristic nightmare of a man walking in the city at night who is stopped by a robotic policeman: His claims that he wants some fresh air (but isn't indoors receiving it from his air conditioner) and is out to see the sights (but isn't indoors watching them on television) get him arrested and carted off to an insane asylum.

situation as The Big Bake. The *New York Times* noted that, during the summer, "cocktail parties went into the bedroom, where the new air conditioner offered some relief." And sales topped a million units.

Summertime heat wasn't the only factor boosting this new popularity. The 1950s were shaping up as a decade of rabid, near-obsessive consumerism: At the same time, they were a decade of strict conformity. Status symbols were becoming excruciatingly important, especially in the close confines of the newly built suburbs with their cookie-cutter houses. There was a strong undercurrent of feeling that if one's home couldn't announce one's upward mobility, one's possessions could—but as opposed to the showoffishness of the 1920s, this version of the game had to be played subtly. To that end, window air conditioners fit the bill perfectly. As big-ticket appliances, they were just about the ultimate status symbols; they were as visible, jutting from a window, as any TV antenna would be on the roof; and, most insidiously, their *whirr* reminded everyone of their presence day and night, all summer long. And while the neighbors might sneer at a flashy Cadillac, a purchase of home cooling could be self-righteously defended as Improving the Family's Comfort.

To ride the crest of this trend—and to combat the aversion that some rock-ribbed souls still felt when they said, "God made bad weather, so you should put up with it"—air conditioner advertising itself changed. Throughout the 1940s, ads had stressed little more than a dry rundown of comfort and technical features. Now, the pseudo-science of Motivational Research had been developed to help advertisers understand why consumers bought, and particularly why they didn't. Through such means as questionnaires ("Complete this sentence: 'If my neighbor installs an air conditioner, he probably bought it in order to _____'"), advertisers discovered that choosing any such unit could be an emotionally loaded decision. So air conditioners were "repositioned," shown as essential components of the upward move to Gracious Living. And as opposed to central cooling, whose blurbs were directed at The Man of the House, room units now were offered in ads intended to appeal to The Little Woman. As advertising director Salvatore Diana put it, "At that time, even though men supposedly made all the decisions, women were the real house managers. And the adoption of the air conditioner reflects the mentality of the time. *She* was the one at home, spending all day in that hot house. *He* was already enjoying it at his office. Those ads gave her the psychological ammunition to justify an air conditioner as a good, sensible purchase."

Not merely the ads but the air conditioners themselves set out to charm female buyers . . . not with capability, but with style. Fresh'nd-Aire offered units in "a choice of color combinations" to match any decor. International Harvester, which had spent more than a century making tractors and other farm equipment, added air conditioners to its line and announced the "Decorator," a unit that could be custom-covered with a swatch of the buyer's own drapery material: "It's a treat to look at even in the Winter!" Remington probably topped them all when it came out with a model that used the "Airflo Fresh'ner," a plastic cartridge that plugged into the unit and dispensed the ultimate miracle-deodorizer fad of the decade, chlorophyll, to sweeten the indoor atmosphere. Even Carrier, no-nonsense granddaddy of the industry, bowed to the trend when it redesigned its room air conditioner, named it the "Silhouette," and posed it opposite high-fashion models (lest men feel slighted, one ad featured a crisply tailored executive in a paneled office). It was no accident that, when Carrier plunged into TV advertising, its spots appeared on two shows with predominantly female audiences: NBC's "Today Show" and the housewife-targeted "Home Show."

These strategies exploded the market until, by 1955, one in twenty-two American households had at least one air conditioner, and a survey conducted by New York power company Consolidated Edison revealed that room air conditioners had "replaced television and the washing machine as the most wanted item by homeowners." Prices dropped in response; now even name-brand units could be found for under $200. Or the really frugal handyman could order a set of plans advertised in *Popular Mechanics:* "Build Your Own AIR CONDITIONER from Junked Refrigerator for Less than $15."

All this enthusiasm was generating a new crop of science-based views of the future. The chairman of the Atomic Energy Commission was predicting that electricity would soon become "too cheap to be metered"—presumably good news to air conditioner owners, as the average unit used eight to ten times the power of a family refrigerator. RCA head General David Sarnoff was predicting (along with an entirely new type of TV set, which "may well be flat and in easel-like frames . . . attached to the walls like pictures") a completely electronic air conditioner. It would be "a noiseless machine without any moving parts."

Willis Carrier wasn't around to comment on these ideas. He had died in late 1950, still active in the Carrier Corporation and still involved to the end with research. Oddly enough, with all his emphasis on the perfectly controlled climate that could be achieved only with central air conditioning, he had never installed it in his own home.

THE NEW SILHOUETTE *Carrier* ROOM AIR CONDITIONER

BUILT BY THE PEOPLE WHO KNOW AIR CONDITIONING BEST. CARRIER CORPORATION, SYRACUSE, NEW YORK

This 1954 ad sells air conditioning with lush photography and *haute couture,* even though it gives not the slightest hint as to the possible connection between the beautifully dressed ladies and the Carrier "Silhouette." Are they admiring its design? Cooling off in the middle of a fashion show? Making a service call? (Courtesy: Carrier)

Perhaps the ultimate example of air conditioning's glamorization. Cabinetmaker Edward Friedrich began building ice-cooled display cases in the 1880s; his sons moved into mechanical refrigeration in the 1920s and entered the home cooling market in the early 1950s. This 1956 ad stresses two vital points—first, the unit's "GOLDENsilence" feature; second, the information that the model's ball gown came from "the Sapphire Room, Russell's, San Antonio." (Courtesy Friedrich Air Conditioning Co.)

Washington's Hot Air (Part VII)

When it came to household improvements, one of the most backward homes in the nation was the White House—a building that had been very hard-used, and not at all expertly maintained, for nearly a century and a half. Signs were beginning to appear that the building was in need of major repair, noted by the Truman family; chandeliers swayed, ceilings sagged, and the President's bathtub seemed inexplicably to be sinking. By the summer of 1948, the problem really announced itself. The President's daughter, Margaret, recalled, "As a matter of fact, the need to repair it became dramatically evident a few days before the campaign started when Dad went upstairs and found that my Steinway piano had fallen through the floor. A large hole had opened up in the second floor and the piano was sitting there tipsily, with one leg in the hole." An inspection was immediately ordered. Washington's Commissioner of Buildings announced that the floor was "staying up there purely from habit." By early 1949, the White House had been "condemned as a structural and fire hazard."

For the next three years, the Trumans lived across the street in Blair House while the White House was gutted, leaving only the outer walls standing, and an exact duplicate of the previous house was built within the shell. There were a few improvements, among them additional basement space that was excavated to give room—finally—to a full-scale central air conditioning system.

The White House was one of the most prominent political symbols that America had ever seen air-conditioned. At the same time, another symbol, and one that had been even more soggily uncomfortable, had been the national political conventions. A badly ventilated hall, in the hottest weeks of summer, stuffed with too many bodies, most of them male, many of them under the influence, all of them determined to put on uproarious displays of enthusiasm that included frenetic shouting matches, occasional fistfights, and hour-long "snake dances" in the aisles—since the first national convention in 1832, it had been a recipe for steaming misery. According to historians, the single upside was the fact that the heat forced delegates to get their business done more quickly.

But aside from passing out palm-leaf fans and trying to administer medical attention to collapsed delegates, little had been done about it.

There had been a few unhappy attempts. In 1916, the Democratic Convention in St. Louis had tried at one point to provide relief for its attendees by blowing air over twelve tons of ice. The nonexistent result was reported by the columnist Irvin S. Cobb in the *Pittsburg Press* under the headline "Smothering Men Talked to Death at Convention": "Inside an air-cooling device . . . was supposed to make bearable the atmosphere, but for some reason the press stands insisted on getting warmer and warmer until finally the limp and gimpless occupants were stewing in their own perspiration."

In 1928, Republicans met in Kansas City, in a hall equipped with two blowers that had been installed right before the convention . . . the 1900 convention, to be exact. At the time, they were much admired. But people's ideas of comfort had changed since those days, and now *The Outlook* snapped, "Convention Hall at Kansas City is not well ventilated, and this was unfortunate. . . ." Democrats that year were no better off; masochistically, they met in Houston, in a temporary wooden structure that was "equipped for comfort" with a number of refrigerated water coolers and two huge ceiling-mounted Typhoon fans—completely open to the sky, a fact that was emphasized when a sudden thunderstorm drenched the delegates sitting beneath them.

In 1932, both parties' conventions were booked into the Chicago Stadium, a venue that claimed to be "air conditioned" but was served only by an air washer. At both conventions, that system's cooling capacity was rapidly steamrollered by the size of the crowds. Inadvertently rubbing salt into the wound, a few reports noted jealously that NBC and CBS radio crews were comfortably ensconced in booths cooled by Frigidaire equipment.

In 1940 Philadelphia hosted the Republicans at the Municipal Auditorium, which a United Press correspondent described as "complete with modern air-conditioning apparatus, which will spare convention delegates the hardship of a Philadelphia June." That wasn't exactly true. The "apparatus" was a fan system that blew air over ice; an estimated forty tons of ice would be needed each hour; all of it would have to be brought to roof level, manually; and no one was prepared to do this. One day, the building staff succeeded in wrestling eight tons of ice up the Auditorium's six stories. It made no difference; 105 delegates were treated for heat exhaustion. Writer Marcia Davenport described the scene as "a filthy, sweaty hell of sealed-in heat."

To make things even more hellish, 1940 was the year that television got involved, as RCA, Philco, and General Electric teamed up with NBC to beam the Republican convention to viewers in Philadelphia, Schenectady, and New York. It was an unprecedented and exciting idea but made difficult by TV's primitive cameras, which required more brutal floodlighting than any film studio. Sure enough, the moment the lights came on in Philadelphia, startled convention delegates began rustling their (elephant-shaped) cardboard fans. Before long, a number of light-blinded conventioneers donned sunglasses. By the end of the first ballot, the delegates were so overheated that the convention chairman ordered the lights be turned off.

When the Democrats met in Chicago a few weeks later, they proved that they had been paying attention. The *New York Times* reported not only that 2,400 tons of ice had been lined up "to counteract the heat generated by the 1,000,000-watt battery of lights" but that "The use of television apparatus, the heat from which added to the discomfort of Republicans at their Philadelphia convention, has been ruled out."

Both parties returned to Chicago for their 1944 conventions, where wartime restrictions meant that there would be no ice to cool the Chicago Stadium (for that matter, one official admitted that past ice-cooling efforts had been "no more than a spit in the ocean"), but more urgent problems than personal comfort were occupying everyone's minds. Once the nation returned to peacetime, however, the 1948 conventions proved to be the last straw.

Republicans and Democrats returned to Philadelphia's Municipal Auditorium—specifically because both parties wanted to try television again, and in the days before TV's coaxial cable had been strung to reach coast-to-coast, an East Coast venue would bring coverage to the greatest number of viewers. For its video appearances, the Auditorium had been spiffed up with $650,000 in improvements, but air conditioning "was rejected as too expensive."

That turned out to be a real mistake. As the GOP convention got underway, the temperature of the hall climbed into the mid-90s, television lighting added its own torrid heat, delegates dubbed the hall "The Steamheated Iron Lung," and the *New York Times* wrote that "the hall gave the appearance of a conclave of Arab sheiks" because delegates were wearing handkerchiefs on their heads "to take up the run-off." When Democrats met a few weeks later, things got even worse. Philadelphia was hit with a genuine mid-July heat wave, temperatures inside the hall topped 100 degrees, more than a hundred delegates were felled by the heat, and the night Harry Truman arrived to deliver his acceptance speech he reportedly got a whiff of the atmosphere, grabbed a chair, and retreated to the sidewalk to wait things out where it was comparatively cooler. Both conventions provided viewers with the spectacle of exhausted delegates, thoroughly disheveled and fanning themselves with whatever was available, along with speakers—including Republican nominee Thomas E. Dewey—who were drenched with perspiration, smarting under the lights, and visibly laboring to stump up energy in the thick heat. The print media had always treated the sweating-politician theme as a genial joke, but when it came to viewing it in close-up on the home screen, everyone realized that it made for vivid news coverage but unappetizing TV. Obviously, if television was going to be a part of political conventions, air conditioning would have to be a part of it, too.

So the 1952 conventions showcased several dramatic changes, mostly due to their new video visibility. As they were going to be covered gavel-to-gavel, they would be held in Chicago, which by that time provided a nicely central location for nationwide telecasting. Both conventions would be held at Chicago's International Amphitheatre, a smaller and less-handily-located venue than the Chicago Stadium but better equipped for television crews. And to seal the deal, the Amphitheatre agreed to install $350,000 worth of Honeywell air conditioning for the occasion. Once the conventions got underway, nearly every news report had something wonderful to say about the newly cooled air. As it turned out, though, the system wasn't able to cool the stage, ablaze with TV lighting. When Massachusetts Governor Paul Dever gave a "blistering" keynote address to the Democrats, the *New York Times* wrote later, he "seemed to millions of viewers to be in imminent danger of melting." In some political circles, "the Dever Wilt" briefly entered the language to describe a heat-sogged speaker.

Rapidly, television and air conditioning became strong influences in the very look of political conventions, and for that matter all of politics. Witness the custom-crafted lectern used in the 1956 Democratic Convention. It contained hidden spotlighting to smooth out facial sags and shadows; it had that astounding new device that made any speaker look confident when reading his speech, a built-in TelePrompTer; the floor was actually a miniature elevator, enabling speakers of every height to be centered perfectly in the TV picture; and to avoid the indignity of the Dever Wilt, the podium was equipped with "three air-conditioning vents" to bathe every speaker in *coolth.*

The Ventilating of Video

It was no secret that air conditioning owed a great debt to television for its popularity—the industry was absolutely delighted at the reams of publicity generated by the introduction of climate control at political conventions. But it was equally true, if less publicized, that television owed its very existence to air conditioning.

Where radio had enjoyed a very short gestation period before it burst onto the public scene, the word "television" had first been used in 1900. Even before that, newspapers and magazines would amaze readers with artists' conceptions of one device or another that was going to "scientifically" bring entertainment, news, and exotic scenery right into everyone's homes; Edison himself had predicted a "telephonoscope," with a gigantic screen, in 1879. But there had never really been any science to back up the drawings. By the mid-1920s, with radio a raging success, television was little more than a theory, but the public was clamoring for news of television.

This might have been the reason that TV's earliest efforts smacked of the laboratory, also of Amateur Night. It was understandable. Television was a much more complicated proposition than radio, trying to transmit not only sound but picture; the machinery itself varied between countries, even between cities, and often changed from one month to the next; and all the while that researchers were trying to solve mind-boggling technical hurdles, there was intense pressure to come up with a result, any result. So the initial efforts, and mistakes, went out over the airwaves for anyone to see.

Few "bought" sets were even in existence. Telecasts were made to be seen by technicians, executives, and occasional invited dignitaries, but actually most viewers were fanatics who had cobbled together their own receivers. Huddling in front of screens as small as a business card, these diehards looked forward to each day's ration, often as little as an hour, of "radio-vision." And every telecast was sternly labeled Experimental. Not that the viewing was all that fascinating; one inventor transmitted an image of a toy windmill, another used the head of a ventriloquist's dummy, and still another parked his wife in front of the camera until she became bored. In 1928, General Electric's television station in Schenectady made the front page of the *New York Times* when it tried to liven things up by televising a short play. The equipment was so clumsy, and the setting so cramped, that the cast numbered two, and cameras could focus only on faces or hands. It was received by an estimated total of four sets. Not long after that, *The Forum and Century* sighed, "To the radio industry, television is not new and not miraculous, but a cat which got out of the bag too soon. . . ."

Insiders might have been amused by that comment, because in fact a cat did figure greatly in early TV—a papier-mâché model of movieland's Felix the Cat, who spent nearly ten years parked in front of RCA experimental cameras, revolving on a turntable for two hours a day. He was a perfect television subject for his time: photogenically detailed in contrasting black-and-white, tireless, and non-union. More to the point, Felix worked under conditions that would have fricasséed an ordinary cat. Television cameras of the time were dismally insensitive to light, which meant that any telecast had to be flooded with more than twice the illumination of a movie. Even the smallest set could be hammered by up to 50,000 watts of incandescence. Along with the heat that it generated.

Even though a TV signal could carry no farther than fifty miles or so, and no one had yet figured how to link television stations into a television network, by the early 1930s both NBC and CBS were confidently forging ahead with the new medium in studio spaces that had been converted to "experimental" TV production. When the networks had first built those studios, air conditioning had been praised as the key to transforming

NBC publicity chuckled, "Felix the Cat has been turned into a grinning *guinea pig* as the subject used most often in the NBC experimental television studio." (New York Public Library, Mid-Manhattan Picture Collection)

radio work into a bearable experience. But as everyone soon discovered, plenty of *coolth* for a radio broadcast was nowhere near enough *coolth* to have the slightest impact on television lighting.

Truth was, anyone stepping before a TV camera was in for a session of unmitigated torture. The press tacitly acknowledged this in just about every article it ran, mentioning television's air-conditioned studios while immediately following up with a comment on the blazing heat. (Things were worse for performers at studios that tried to get along without any air conditioning at all, such as the one operated by low-budget DuMont out of a Manhattan office building, where temperatures could top 140 degrees.) And while articles tried to stress the "strong interest" that TV was getting from show business luminaries, those articles lied; if well-known personalities tried the medium, one experience was usually enough. Mary Pickford appeared once; although she was one of film-dom's most seasoned veterans and had her own horror stories of trying to act under blazing Technicolor lights, TV's studio temperature aston-ished her: "This is hotter than *color*!" Ex–New York Governor and one-time presidential candidate Al Smith, himself no stranger to bright lights or overheated rooms, was asked to take part in a short talk on CBS station W2XAB. When he came out of the studio, perspiration rolling down his face, he gasped that television would never make it until there was some way of "cooling off that place in there." And when Broadway superstar Gertrude Lawrence agreed in 1938 to do a snippet of her play *Susan and God* for NBC's W2XBS, she walked into an environment where camera-men were wearing pith helmets, and musicians sunglasses, to shield themselves from the wattage. By the end of the half-hour telecast, one reviewer commented that she "looked quite tired, undoubtedly caused by the intense heat." For years after, she gave TV a wide berth.

The technical journal *RCA Review* had listed air conditioning as one of the "sciences" that would be absolutely vital if TV were to succeed as a business. In Paris, station Radio-PTT Vision took this advice seriously. With lighting that was described by an alarmed observer as "the glaring arc of an electric welder," the station kept performers upright with the help of six nautical-type air funnels, just out of camera range, that blasted chilled air directly at them (with enough force that the *Hartford Courant* reported one lady's skirt being blown over her head during a telecast). Elsewhere, the message wasn't getting through. At stations around the world, stories abounded of candles that bent double in their holders before they could be lit, of violin strings that popped in the middle of a note, of beverages in a glass that suddenly began to boil. A science pro-gram was intended to show viewers the microscopic life in a drop of water; as soon as the lights were turned on, every such organism died. *Billboard* described the furniture on the set of one show which became so hot under the lights that it raised blisters on an announcer's backside. Worst of all were food demonstrations. In more than a few instances, the dish in question instantly turned rancid under the scorching lights, and if an announcer was scripted to chew, swallow, and praise, technicians learned—the hard way—to have a bucket waiting just out of camera range.

The problem still hadn't been solved when NBC, which had plunged nearly $50,000,000 into TV's development, decided that, ready or not,

During this telecast for Radio–PTT Vision, the singer's incandescent performance was made more so by twenty high-intensity lamps. Some performances used even more light. (*L'Illustration*)

television would be "officially" and "commercially" unveiled to the public at the 1939 World's Fair. It was a sputtering start, with plenty of press coverage but a grand total of no more than "possibly 1,000 persons" watching the telecast on W2XBS. And anyone who bought RCA's top-of-the-line $600 television set, offering an extra-large (12″) picture in a veneered cabinet, was in for disappointment. Top-rank talent still preferred radio work to the sweaty discomforts of TV, and interest was sporadic enough that NBC couldn't completely fill a full-time schedule. The *coup de grâce* came with America's entry into World War II, when the television industry was virtually suspended for the duration.

After hostilities ended, TV came back to life with a roar and only got bigger, helped along by wartime technology, a brand-new network system, and a public that was more than ready for video entertainment. RCA helped, too, unveiling a camera that needed only one-tenth of the usual lighting and could supposedly pick up a face by candlelight (ads crowed, "No more blazing lights!"). Theoretically, telecasts would be cooler. But as it turned out, not that much cooler. "One-tenth of the lighting" was still a lot of lighting. Lucille Ball made her first-ever TV appearance as a guest on DuMont's game show "Charade Quiz." Even though Lucy was a trouper who had spent more than a decade on film sets, DuMont's hot lights got to her; only minutes into the telecast, she was sopping wet, furious . . . and vowing that she would never again appear on television.

While Lucy would ultimately eat her words, and more profitably than any other performer in TV history, at the time her complaints didn't make the front page. Far different was the brouhaha that developed in the spring of 1948 when legendary conductor Arturo Toscanini agreed to let the NBC Orchestra, a radio staple, be seen on television in an hour-long program. It was a truly historical occasion; a very large chunk of the viewing public tuned in to the program, along with *Life*, which published mid-performance photos of the eighty-one-year-old maestro, in a black wool suit and high collar, "sweating with the heat of the lights" and seemingly transported by the music. But as it turned out, Toscanini, along with a number of his musicians, hadn't been enraptured as much as nearly toppled by the heat. With a televised Beethoven's *Ninth* coming up in a couple of weeks, the maestro shocked network executives, along with the press and the rest of the universe, when he announced that he *would not* conduct that program for TV cameras. This announcement produced days of highly publicized back-and-forth. Finally, Toscanini changed his mind—only, however, after NBC brass guaranteed that Studio 8H would be thoroughly chilled by running the air conditioning for three hours before the performance. *Refrigerating Engineering*, which didn't get to write about symphonic music very often, reported that "NBC officials scurried about, told their air conditioning engineers to give the Maestro a cold blast, promised him that the atmosphere would be to his liking if he would relent. . . . Refrigeration can make a studio comfortable for any length of program under a blaze of lights if properly installed and applied."

Indeed, this would become the industry's standard solution to the problem of heat—*more* air conditioning, running colder, operating up to twenty-four hours a day, and transforming studios into iceboxes. (This in itself would become a fascinating behind-the-scenes part of studio tours. A former NBC page told of groups' being brought into the empty studio of a well-known game show, the visitors' surprise at being able to see their breath, someone invariably asking why, and his pleasantly deadpan response, "This way, [name of a certain overripe TV glamour queen and eternal game show panelist] keeps fresh between shows." It never failed to get a yelp of laughter.)

But this solution was an imperfect one, and it would always have to be adjusted. When CBS put together its first "experimental" color telecast in 1951, it was a nationally covered event, but in some ways it recalled TV's first faltering steps of the 1920s. The network had pulled together a high-profile cast for visibility . . . the show was viewable only on special sets equipped to receive it . . . and the production was nearly sabotaged, as CBS color cameras required lighting so much hotter than anything needed by black-and-white video that it overpowered the air conditioning. Studio linoleum buckled and lifted off the floor. As soon as a sponsor's pie had its close-up, it collapsed from the heat and the cherries rolled out of the crust. On camera, normally sedate emcee Arthur Godfrey pulled out a handkerchief to dab his forehead, grumbling about the temperature. Of course, *Life* was right there, giving readers not only color images of the program but also a candid shot that showed a studio cameraman and ballerina Tanaquil LeClerc, both trying desperately to mop off in mid-show.

And there would still be those occasional moments when an old-fashioned remedy was needed. A CBS technician remembered a Friday night in 1953 when the studio air conditioning failed but an episode of "Mama" was due to air that evening. Someone ran out to buy washtubs, electric fans, and a great deal of ice. And the show went on.

Towers of Coolness, Continued

As homes and entertainment changed in this era, so did the very look of the urban landscape, and it was largely due to climate control. Some people were taken aback by this. They shouldn't have been shocked; the architectural critic Lewis Mumford had given ample warning years before when he wrote that there was definitely a "façade demanded by air-conditioning." The postwar building boom was going to take full, and occasionally unpleasant, advantage of it.

Once the PSFS Building transformed the Philadelphia skyline in 1932, the world of architecture had been rocked to the core by the form-following-function concept of International Style . . . which, to many architects, was only a starting point. Since the early 1920s and even before, many designers had been panting for the aesthetic ideal of a building with all-glass walls. Back then, the simple difficulty of getting uniform large sheets of glass made this kind of construction an expensive and unrealistic notion; but in the postwar era, improvements in glass manufacturing made it possible to fulfill the fantasy. Architects, and a surprising number of clients, were raring to give it a try.

If the idea of a climate-controlled tower of steel and glass had once belonged only to science fiction writers and sophisticates, the extent to which it had now become thoroughly absorbed into mainstream culture was evident from the fact that the first one built in the United States arose not in a major metropolis but in Portland, Oregon. In 1948, Equitable Savings and Loan moved into its new home, a twelve-story building that *Architectural Forum* praised as "a long overdue crystal and metal tower." While it wasn't a headline-maker, it incorporated a number of "innovations"—or, more accurately, they weren't innovations, but for the first time all of them were combined in one building. The Equitable Building's outer shell was aluminum and glass (but there had been other "glass-box" buildings). It was completely air conditioned (but there were the Milam Building and its brethren). And—shades of the turn-of-the-century Armour Building, the Larkin Building, and even the Chicago Public Library (but possibly no one remembered such long-ago examples)—it was constructed with sealed windows. The architectural establishment was excited by the result, and a precedent was set.

At the same time the Equitable Building was rising, across the continent the newly formed United Nations was planning its headquarters in New York. To come up with a design, the job was put into the hands of an international ten-member team of architects, among them Le Corbusier. In the years since the overheated debacle of his Cité de Refuge, he had given up on *l'air ponctuel* for temperature control and had concluded that buildings should have operable windows. In addition, he had become known for buildings with *brise-soleil*, "sunbreakers," that functioned like louvers and shielded sun-facing windows to provide natural heat control. Because "Corbu" knew New York's sizzling summer weather, and because the *brise-soleil* had worked out well in the design for a government building in Rio de Janeiro, he recommended them for the UN's centerpiece Secretariat Building. It would certainly need them, as plans called for two of its four walls to be composed entirely of glass. He wrote to the head of the UN Advisory Committee, "My strong belief is that it is senseless to build in New York, where the climate is terrible in summer, large areas of glass which are not equipped with a 'brise-soleil.' I say this is dangerous, very seriously dangerous." But the UN Planning Office, and the rest of the design team, preferred the more up-to-date quality of air conditioning. The *brise-soleil* were overruled. All that Corbu got of his ventilating scheme was windows that opened.

The Secretariat was an eye-opener, thirty-nine stories of blue-green Thermopane "heat absorbing" glass; but as soon as the building was

The United Nations Secretariat Building, characteristically (and mercilessly) illuminated by the sun. (Library of Congress, Historic American Buildings Survey)

ready in August 1950 and staff moved in, it became evident that the glass, if it absorbed heat, didn't absorb enough of it. Moreover, leaving off the *brise-soleil* had been a definite mistake. Even with a system of 4,000 Carrier units operating throughout the building, west-facing offices received a continuous direct hit from the sun's glare and had to spend much of the day with blinds completely drawn. This scandalized Le Corbusier, who inveighed against the "morguish light" and the "sepulchral repression" that the building's workers were undoubtedly experiencing. For that matter, the architectural elder statesman and critic Henry-Russell Hitchcock added his own negative review when he wrote that the building would serve mainly "to end the use of glass walls in skyscrapers—certainly those with Western exposures—unless exterior elements are provided to keep the sun off the glass."

Hitchcock's prediction turned out to be dead wrong; to many minds, if two walls of glass were good, four would be better. This was dramatically achieved in 1952 when Lever House—as the New York Times gushed, "the newest and glassiest of all"—rose on Park Avenue. Hailed as a design triumph by critics and public alike, with an enthusiasm not seen since the opening of the Empire State Building, the very modern structure was welcomed even on a block of decorous stone apartment houses, and soap manufacturer Lever Brothers took out a full-page ad in the Times merely to tell an entranced populace about the features of its new home. Its slim twenty-four-story tower of "sea-colored" glass made it a very small building by New York standards. But its impact was gigantic.

Even more important than its appearance was the fact that it justified all those old science fiction stories by offering a totally controlled working environment. Lever House allowed employees to avoid all contact with the grubbier aspects of the city; they could drive directly into the building's underground garage, take meals in the company cafeteria, and work in its well-lighted offices. As the ultimate touch, the building was constructed with non-opening windows "to seal out dirt and dust," and every cubic inch of it was air-conditioned. One observer noted acerbically that Lever employees didn't even have to breathe the same air as other New Yorkers.

Almost at once, Lever House became the new standard, its features copied and "interpreted" in office buildings the world over. And air conditioning was largely responsible, not only for the operation, but the very look of these glass towers. Because windows were no longer needed for ventilation or sunlight, the old rules of skyscraper architecture became obsolete; windowless "deep space" was now perfectly acceptable housing for support staff and even middle management. Naturally, this meant that modern office buildings would take maximum advantage of their sites. Skylines the world over would be gradually dominated by structures that were significantly boxier and chunkier than the slim, tapering towers of decades past.

Lever House was considered a trailblazer on Park Avenue, but at the time no one mentioned that it wasn't the first air-conditioned building in the neighborhood. Immediately to its right in the photo is 400 Park Avenue, the apartment house that had received central cooling as part of a 1936 makeover. Ironically, that building would be razed in 1955 to make way for another air-conditioned steel-and-glass office tower. (Milstein Division of United States History, Local History & Genealogy, The New York Public Library, Astor, Lenox and Tilden Foundations)

For that matter, air conditioning was the reason that these buildings had to find a new way to clean all their glass. Without windows that could open, window washers couldn't climb out onto ledges, which meant that they would need some method of washing windows from the out-side—which led to Lever's installing the very first mechanized "window-washing gondola," suspended from the roof, that traveled around the building and allowed two men to scrub the entire glass exterior every six days—using, of course, Lever's own household detergent Surf. That first gondola created traffic snarls from rubbernecking cab drivers and stopped pedestrians in their tracks. But within a few years, the sight was utterly commonplace in just about any city. So were the buildings from which the gondolas dangled: glass-walled, hermetically sealed, and unvaryingly climate-controlled.

In the meantime, older buildings weren't to be outdone. As Carrier had noticed, there seemed to be a formula to it; whenever 20 percent of the office buildings in any particular city included air conditioning, the other buildings would be forced to include it "to maintain their first class status," not to mention their rentals. Even big-city skyscraper icons took part in the scramble. The Empire State Building was fitted out with air conditioning (supplied only to those tenants willing to pay for it) in early 1951; the Woolworth Building had one-third of its office space air-conditioned in 1953; and the Chrysler Building got cooling—naturally, a Chrysler Airtemp system—the following year. At the same time, *Architectural Forum* headlined, "AIR CONDITIONING—A MAJOR MODERNIZATION TOOL/ Available at $2 to $7 per sq. ft., it helps put old buildings in competition with new ones."*

This new breed of workplace, with its fluorescent lighting, conscious design, customized furnishings, and "modernized" climate control, was a far cry from the office building of even a decade before. Back then, even

---

*It wasn't only for comfort that conditioned air was absolutely vital to the new breed of building; the 1950s were the decade in which computers first made their mark on business life. Contraptions with acronymous names like UNIVAC, SEAC, RAMAC, ENIAC (and even MANIAC) were "genius machines" that functioned with amazing speed, but they were also gargantuan devices that filled entire rooms, used thousands of vacuum tubes and miles of wiring, and in the process generated fright-ening amounts of heat. Any corporation installing computer equipment was instructed to keep its electronics healthy by installing extra air conditioning—a great deal of it—to chill as well as to filter the air. Within only a few years the cliche of the nearly freezing computer room, with shivering technicians and programmers don-ning sweaters even in August, became familiar to every company.

the most famous skyscrapers reserved their luxury for exteriors and lobbies, office floors themselves were bare and unwelcoming, and this was just something that everyone expected. Now, employers had thoroughly digested the fact that comfortable employees were *productive* employees—and every air conditioning firm had access to multiple studies supporting this claim. By the start of the next decade, *Newsweek* would report that it was simply no longer possible to rent "Class A" office space that didn't include air conditioning, and even the Australian business magazine *The Valuer* chimed in, "It is not considered wise, or even usual, to erect any office building which is not air conditioned."

But a sinister-minded observer might insist that this building-wide comfort came at the price of tenant freedom. Back in 1949, management at the pioneering PSFS Building had decided that its thousands of windows, which originally could be opened by tenants for a breath of air, interfered with uniform control of the air conditioning. So they were locked.

Cool Conveyance

In 1959, Carrier chairman Cloud Wampler said, "After fifteen extremely painful years of adolescence, I think air conditioning is at last ready to come of age."

In any case, by the 1960s it seemed that air conditioning was able to go anywhere. There was the vending machine used in Texas drive-in theaters that dispensed two hours of conditioned air for a quarter; the air-conditioned space suit worn by NASA astronaut Alan Shepard when he piloted the *Freedom 7* rocket; the six immense truck-mounted units that cooled the audience under the big top of the Barnum & Bailey Circus; and in the Shreveport (Louisiana) Air Terminal, there was a test installation of the Rest-O-Booth, a small wooden cubicle that rented for twenty-five cents per half hour and provided frazzled travelers with privacy, air conditioning, and a reclining "automatic contour chair"—"automatic" because the seat was designed to announce the end of the traveler's nap time by sproinging back into upright position. At the outer edge of probability, in 1960 Buckminster Fuller echoed John Gorrie's fantastical predictions of the 1840s when he outlined plans for a geodesic dome, two miles in diameter and a mile in height, that would cover Manhattan's Midtown district to "regulate weather and reduce air pollution." Weather inside the dome would be controlled year-round, buildings would need no air conditioning of their own, and Fuller predicted that the savings in the cost of snow removal alone would pay for the entire project in ten years. Somehow the dome was never built.

As far as members of the public were concerned, it was not only nonsensical but infuriating that engineers could keep a straight face when discussing the prospect of air-conditioning a whole city, while insisting that it was impossible to cool rapid transit. This had been a universal problem, with the most publicized offender being New York; the subway had expanded steadily through the 1930s until it became one of the largest systems in the world, but visitors and old-timers alike found it nauseatingly hot (and the open windows and doors in subway cars made for a deafening ride as they hurtled through the tunnels, not to mention a dirty one). In 1946 the chairman of the Board of Transportation told the *New York Times* that "Air-conditioning cars with four doors opening every few minutes is out of the question, and even effective ventilation is a difficult problem"—a proclamation that was received with raised eyebrows by the public, considering that Atlanta had only the year before installed the first of a fleet of trackless trolleys, equipped with full air conditioning.

The New York system's reputation got no better in the following years. In 1948, the old ceiling fans in subway cars were replaced with "an adaptation of larger ones used during the war to ventilate the hulls of Navy ships." As this didn't provide much of a trade-up in comfort, in 1954 the system tested other fans, supposedly controlled by thermostats to make them turn faster in high temperatures and designed to "draw in air from the subway tunnels, fresher and cooler than car air."

By 1955, things were bad enough that the system was experiencing dwindling ridership; coincidentally, that summer saw a trial of a single air-conditioned subway car. As the equipment was severely underpowered for what it was being asked to do, it worked when the car was half-filled with dignitaries and reporters but failed miserably the very next day when it tried to cool the usual sardine-packed load of commuters. Another system was tested the following summer, along with Muzak in subway cars; neither the cooling nor the music worked out. Transit officials were further embarrassed when the subway's New Jersey–to–New York rival, the Hudson & Manhattan Railroad, not only tested its own air-conditioned car at the same time but added nearly fifty of them to its fleet in 1958. Frustrated, in 1962 the chairman of the Transit Authority announced that the $300,000 "experiment" had come to an end.

Apart from public annoyance, nothing much more was said officially until the 1965 New York race for mayor, when candidate John Lindsay brought the house down at a campaign event by promising that, under his administration, the subway system would get air conditioning. He took steps to make good on his promise in 1967, when four cars were test-equipped with new and improved systems. Riders were initially skeptical, but then enthusiastic—except for one woman who saw two engineers measuring the results with foreign-looking instruments and ran to a trainman to report that someone was setting off a bomb—and city residents eagerly looked forward to cool commuting. But Lindsay's prediction, that the entire system would be air-conditioned by 1980, was premature. Given the effort and expense it took to replace or retrofit cars, it turned out that a New York subway passenger wouldn't be assured of riding in an air-conditioned train until well into the 1990s.

In comparison, it took far less time for air conditioning to establish itself as a necessity in most automobiles. Timing was part of it. Everyone had been so car-famished during World War II, and so delighted when postwar production started up again, that members of the public greeted the new models with a rush of interest. So did the press. And a great deal of interest was centered on new and exciting features. *The Rotarian* reported in 1946 that new cars were routinely going to be air-conditioned. And stalwart *Life* reported on a poll taken by the Society for Automotive Engineers, in which respondents in major cities had been asked for their wish-list of automobile features. Air conditioning made the list, particularly in New Orleans.

But when it came to turning the wish into reality, there were snags. Packard may have started the whole thing, but it had given up on the Weather-Conditioner in 1942 and made absolutely no move to offer it again. Nash, with its non–air conditioning Conditioned Air System, had in 1939 added a thermostat to the heater and renamed the system the Weather-Eye, promising, "Don't say it's too cold. We will spin a dial and conjure up a warm May day." But Nash was careful to make no claims that the Weather-Eye had any way to *cool* the air flowing through it. Other manufacturers in the postwar market weren't so fastidious; a number of them, from Kaiser-Frazer and Jaguar to Rolls-Royce, were offering similar systems and casually labeling them "air conditioning" without further explanation. The British journal *The Autocar* became irritated enough by this type of fib to write in 1949, "It is unfortunate that the term 'air-conditioning' has been used too soon in connection with motor vehicles. The heaters and ventilation systems now in use add much to motoring comfort, but the fully air-conditioned car is still far in the future."

For a few years, Detroit continued to tinker with the problem, while motorists resorted to their own ingenuity. The "car cooler" became popular, especially in western states; a portable version of the swamp cooler, it hooked over the front passenger window, scooping up air, blowing it through a moistened pad, and routing it into the car. A motorist could rough it, banging together his own car cooler from plans in *Popular Mechanics*, or splurge and order the Thermador Car-Cooler, finished to match his car's paint job. Southwestern travelers could even rent car coolers, picking them up from stores at one edge of the desert and returning them at the other side. For less efficient but more portable cooling, most gas station chains handed out free cardboard fans; and some Continental Oil stations offered a curbside Airtemp unit, connected to a long hose, that would pump "gobs of invigorating cool, dry air" into a customer's car, enough to lower the temperature by twenty degrees. (Passengers were told that with windows rolled up tightly, they'd be comfortable all the way to "the next stop," wherever that might be.) Some people thought to travel with chunks of dry ice, suspended in towel slings from the inside windows of the car—again a potentially lethal idea, as the release of all that carbon dioxide could cause everyone in the car to black out right on the road. Less dangerously, there was a small rubber-bladed fan that plugged into the cigarette lighter and clipped on to the dashboard to give the driver a breeze. Failing any of these remedies, drivers joked about the least expensive cooling system of all: "Four-Forty Air Conditioning—four windows down, forty miles an hour."

As all of these ideas had flaws, there sprang up a handful of companies that specialized in installing "after-market" air conditioning. As this wasn't a bit cheap, it was considered by most people to be quite a raffish move, only for VIPs. In 1951, the *New York Times* reported that the government of Argentina had ordered an extraordinarily customized Cadillac for Juan and Eva Perón. It came in at $11,000, a perfectly reasonable amount considering that it had been equipped with a rear-seat TV screen, built-in bar, mobile phones, folding tables, bulletproof glass—and air conditioning by the Kool-Kar Company.

It wasn't until the summer of 1952 that General Motors made the announcement that 1953 Cadillacs and Oldsmobiles would be offered with "refrigerated air conditioning," developed of course by GM's Frigidaire division. Within months Chrysler, and then Packard, joined GM. In each case, the system was very much a descendant of the 1939 Packard machinery, taking up trunk space, requiring the driver to climb into the back seat to make adjustments because some controls were mounted

Automotive Digest endorsed the Thermador Car Cooler as "an accessory designed to make hot weather driving a pleasure." But one driver wasn't as enthusiastic: "Don't work a damn in humid areas." (Photo: Doug Coldwell)

there, and adding 250 pounds to the car's weight—as well as nearly $600 to the cost, which in 1953 dollars made it just as expensive as Packard's original Weather-Conditioner had been. Still, by the following summer Chrysler reported that demand for air-conditioned cars was outstripping its production capability. Customers might have wished they had waited. *Popular Mechanics* reviewed the new air-conditioned Cadillac with less than wild enthusiasm: "GM's air-conditioning system is somewhat complicated. . . . The body seems to be honeycombed with ducts, scoops and vents . . . It is a little noisy when the blowers are on high." And there were scandalized whispers in 1954 that First Lady Mamie Eisenhower, sitting in an air-conditioned Cadillac, had been *dripped upon* by condensation from the cold air output.

If the high price of automotive air conditioning was a barrier, it was shattered the following year when *Financial World* wrote, "Nash Motors claims to have a triple-threat which will 'obsolete' air cooling units now available for automobiles and 'bomb-shell' the industry with respect to price—this is how the company heralds its new All-Weather Eye. . . ." And Nash was absolutely right. The All-Weather Eye was the prototype for the modern car air conditioner—machinery mounted with the engine, *all* controls mounted on the dashboard, more than a hundred pounds lighter than the GM system. Best of all, it definitely came at a bombshell price, $395. Within a year, nearly every auto manufacturer was offering its own version of the All-Weather Eye, even GM, and even on lower-end cars.

Again, the timing was perfect. *Newsweek* was calling the automobile "America's traveling living room," and the Associated Press wrote, "This decade may well go into automotive history as the 'accessory age.'" Next to such fripperies as custom fur seat covers, motorized radio aerials, and tissue holders, automobile climate control made complete sense to a lot of people. *The New Yorker* saw the trend with a curmudgeonly eye, claiming that it was part of an "attempt to reduce all the air in the country to the even coolness of the grave." Still, an air-conditioning installer was quoted: "Are washing machines luxury items? Television sets? For that matter, cars? Why, in five or ten years an air conditioner will be as indispensable as a heater is now." And as a topper, the article noted New York's short-lived experiment with a fleet of fifty air-conditioned taxicabs, clearly marked with a windshield sticker, and extremely popular with riders. One driver happily remarked that many of his fares were people wanting "just to be driven around a while, to cool off."

Industry numbers varied, but their general direction unmistakably told the story. In 1955, 10 percent of American cars were equipped with air conditioning; in 1960, 20 percent; in 1965, 23 percent; in 1968, 40 percent. That last year, 1968, American Motors rocked the industry by offering air conditioning as standard equipment on its Ambassador—at the time, the only other major company that did was Rolls-Royce. Other manufacturers quickly followed suit. The industry, as well as millions of drivers, never looked back.

Air conditioning provided drivers not only automotive comfort but also status. The magazine *Executive* tried to make it a matter of efficiency when it pointed out, "Look what happens to salesmen out on the road. By 3:30 they're hot, they're sticky, their shirts are rumpled, they've had it. But put 'em in an air-conditioned car, they're still proud, their ego's still there, and they make those two or three extra [sales] calls." *Business Week* merely commented, "Tight-closed windows under a blazing sun are the conspicuous cachet of the air-conditioned automobile."

To that end, stories circulated that hard-up Texans without air conditioning were trying to bluff their neighbors by driving with windows rolled up, no matter how hot the weather.

7 The Unnecessary, Unhealthy Luxury (That No One Would Give Up)

The U.S. government gave an unintended endorsement of air conditioning in 1959 when the U.S. Weather Bureau announced the creation of its Discomfort Index, a calculation of heat-and-humidity that was meant to provide an easy guide to the climates of various American cities. In response, Chambers of Commerce in a number of locations objected strenuously to the term "Discomfort"; that word could scare off tourists. Almost immediately, it was renamed the Temperature-Humidity Index. As that was too long a name for the public tongue, it soon abbreviated itself to the pure-and-simple Heat Index. Whatever it was called, members of the air conditioning industry could be (privately) overjoyed. It put people in mind of cooled air.

The industry could be delighted for another reason—if the universal adoption of air conditioning had been a battle, by the 1960s one could say that it had been won. To a great extent, people had been trained to expect it; for the first time in history, if John Q. Public turned into a doorway from the hot street and found that the air inside wasn't cool, he would very likely wonder why. And rightfully so. Now, it was a rare public building that wasn't cooled in summer. Places of amusement found that air conditioning was essential if they were to stay competitive, and so did major businesses. Even small-town stores were displaying the front-door decal that the Brown and Williamson Tobacco Company had first offered to air-conditioned shops in August 1953, featuring the Kool Cigarettes penguin along with a cheery invitation: "Come in—It's KOOL Inside."*

*On the other hand, the decade would see the slow demise of an institution—the frozen candy bar. As chocolate had been virtually unsaleable in hot weather, the notion of a frozen "summertime treat" had been dreamed up by manufacturers, as soon as freezers made it into stores, as something of a Hail Mary marketing move. Even though writer Joel M. Vance remembered the frozen Milky Way as having the consistency of "a sweetened bar of iron," the public enthusiastically responded, keeping candy companies humming through the hot months . . . until store air condition-

A steadily rising number of householders went along with the trend. Some of them might have been helped by air conditioning's Hollywoodization in films like *The Seven Year Itch*, in which a man's attraction to neighbor Marilyn Monroe is matched only by *her* attraction to his air conditioner. The 1960 census showed that close to 13 percent of American homes had air conditioning in at least one room, with the number climbing all the time (it would jump to nearly 37 percent by 1970). Even cars were catching up. And there was still another indicator, more casual but nonetheless powerful; the abbreviation "a/c," once a technicians' term, was entering the national vocabulary. Anyone in the business could say it was a foregone conclusion—conditioned air was making the whole world happy.

But there were detractors, and they had existed when air was first being cooled. Steele MacKaye, proud of his ice-and-fan system at the Madison Square Theatre back in 1880, had received the occasional complaint from audience members who found the theater too chilly for their liking. Alfred Wolff, fine-tuning the refrigeration equipment at the New York Stock Exchange in 1903, would be told that too-cool temperatures could kill off the city's stockbrokers. Ever since the 1920s, there probably had been no movie theater, anywhere, whose air conditioning hadn't come in for criticism: A story had it that one long-ago exhibitor, anxious to prove that his theater was indeed a frosty retreat in hot weather, proudly ran a full-page ad consisting of nothing but angry letters from patrons who had caught colds from his cooling system.

A major reason for such clashes was *comfort*—which, no matter what any engineer said, was very much an individual feeling. It was an undeniable fact that air conditioning had made some uninhabitably hot places wonderfully comfortable. It was also undeniable that the technology had been abused by many users, making some of those places colder in the summer than they ever might be in the dead of winter (all that icy "advertising air" spilling out the doors of those movie houses . . .). The spread of conditioned air had forced temperature-sensitive people to carry woollies with them in July. Many of them didn't like it.

And, as was still the case for a number of puritanical souls, air conditioning was *unnatural*. It even looked strange, with machinery that often made it conspicuous. New buildings were crowned not with elegant spires but with cooling towers. Older buildings, even some of the most

---

ing made frigid chocolate obsolete. Ever after, sentimental summertime gourmets would be forced to freeze their own.

architecturally notable, were disfigured by rows of hastily-plopped-in window units. This was understandable; sometimes the estimate to cool a building centrally was higher than its initial construction cost. One 1920s-vintage courthouse in New York boasted a painstakingly detailed Art Deco lobby. Its window had been casually chopped out to hold *two* air conditioners, one balanced a bit crookedly above the other.

Absolutely, this would create some bad PR. People who really despised mechanical cooling could point to the title of Henry Miller's 1945 harangue on the soulless mechanized commercialism of America, *The Air-Conditioned Nightmare*—inaccurately; other than its title, the book never did get around to discussing the subject of air conditioning. (And Marcia Ackermann pointed out that, once Miller's writings took off, he moved to a large California house that "almost certainly" had air conditioning.) Nevertheless, Miller's bad-boy intellectualism was shared by *The New Yorker*, which throughout its history seemed to have viewed air conditioning with amused disdain. This implied scorn finally crystallized into open hostility in the summer of 1959:

> How useless now to argue that in the old days electric fans blowing across cakes of ice kept saloons at a temperature that was one of the pleasures of summer instead of a simulacrum of winter. . . . The dodges for coping with the heat that New Yorkers learned in three centuries of summers have become superfluous, and in some cases hazardous. The long drink is an irrelevancy; if you arrive in a bar, after a few steps in the street, longing for a Tom Collins, half a minute of the temperature inside influences you to change to a hot toddy. Cold foods lose their charm as quickly; at the first blast of frozen air, the customer decides to stick to steak. . . . It was a fine city until they started improving it.[1]

An impassioned indictment, but a sharp-eyed reader might have found it laughable considering that the rest of the issue featured an advertisement for Carrier (the copy cooed, "We'll take the basement apartment. Carrier can air condition *anything!*"), along with other ads scattered throughout that stressed the air-conditioned comfort waiting at the Roseland Ballroom; the Waldorf-Astoria, Pierre, St. Regis and Plaza hotels; the Beau Soir and Maison Pepi restaurants; and five different Broadway theaters.

Not that *The New Yorker* was alone. Whether as a result of genuine dislike or natural human perversity, by the time that air conditioning was becoming ubiquitous, plenty of observers had decided to hate it, and to somehow link it with the place where it was usually found, the office building, as a dark symbol of The Decline of the City. *Life* now complained of "unimaginative boxes of air-conditioned office space which increasingly dominate U.S. urban architecture" and sneered at those boxes as "air-conditioned merry-go-rounds." The Royal Architectural Institute of Canada condemned "the sterile ideal of the completely air conditioned box." The *American Machinist* wrote, "People tend to go a bit stir crazy in today's highly touted sealed-in (all air-conditioned, no windows that open) office buildings." And a conference involving Columbia University, along with the American Institute of Architects, described a scenario in which

> it would be possible to go from an air-conditioned train to air-conditioned office and back again without coming into contact with ordinary unprocessed air. . . . [We might] yawn our way through a totally artificial life within the city proper, with the air-conditioned taxi taking the place of the air-conditioned commuter train. We might, that is, if there were more to the notion that we have succeeded in producing ideal conditions indoors. As a matter of fact, we have not. The sealed office building is a pretty good place to work, better than an old-fashioned one most of the time, especially under urban conditions of dirt and noise. But anyone who has worked in such a building soon learns that it is far from ideal.[2]

Even Frank Lloyd Wright—who had spent decades claiming that he had started the whole thing with his completely sealed and air-conditioned Larkin Building and had then gone even further with his completely sealed and air-conditioned Johnson Wax Administration Building—blithely and unapologetically reversed himself. In his 1954 book *The Natural House*, he stated, "To me air conditioning is a dangerous circumstance. The extreme changes in temperature that tear down a building also tear down the human body. . . . I can sit in my shirt sleeves at eighty degrees, or seventy-five, and be cool; then go outside to 118 degrees, take a guarded breath or two around and soon get accustomed to the change." Then there was the day he sat down for a drink at the Plaza Hotel with a *New Yorker* reporter, found the atmosphere unpleasant, and berated the waiter for the temperature: "Air-conditioning! Too cold! You know, this air-conditioning has killed more good men than can be accounted for."

The most interesting thing about such potshots was that the general public ignored them. It was all well and good to complain about the "artificiality" of air conditioning if one had the financial resources to skip town when it became too hot, but the average citizen, who had to work throughout the summer, and whose bedroom received no nighttime breeze, could not have cared less about Mr. Wright's directions for learning to love the heat. And the comfort provided by a $150 air conditioner was just too logical to pass up. *Popular Science* spoke for many people in 1966 when it wrote, "Home air conditioning, once a sure sign of plush living, has suddenly become a way of life for practically everybody."

Not only homes were affected by this new standard of comfort. As early as 1957, Manhattan's Lincoln Center for the Arts had issued a flat confession of the outmoded state of the Metropolitan Opera House— interestingly, leading off not with the lack of space or the antiquated stage facilities but with the futility of trying to make its air comfortable. "The Metropolitan Opera House, although nostalgic, is obsolete. It cannot be air-conditioned, hence its economic use is limited. . . ." Sure enough, by 1966 there was a new, air-conditioned Met, joining the other new, air-conditioned performance spaces at Lincoln Center. And the Old Met was rubble.

When even the new elephant house at the London Zoo was climate-controlled, it was probably no surprise that the cooling of the 1964–65 World's Fair—the most extensively air-conditioned fair ever, down to the cars of the Monorail—was no longer newsworthy. (A more appropriate sign of the times might have been the discovery, just before the Fair opened, that a motor belonging to one of the Maryland Pavilion's air conditioning units had been stolen.) Rather than the Fair, the public was fascinated by the 1965 opening of the Harris County Domed Stadium, otherwise known as the Houston Astrodome. Houston had once been touted as "one of the most air-conditioned cities in the world," even boasting a network of air-conditioned tunnels connecting major buildings; now that slogan was spectacularly demonstrated in the world's first air-conditioned baseball diamond. The eighteen-story-tall structure was equipped not only with a transparent roof but with a system that would bring in 250,000 cubic feet of fresh air each minute, cool it to 72 degrees, and keep it there, thereby setting a record for "the world's largest air-conditioned room." If sport itself wasn't revolutionized by the Astrodome, attending games was. Even though *Sports Illustrated* worried that fans would be "coddled" by such surroundings, domed stadiums began to spring up all over the world.*

---

*Even Carrier would get involved, teaming with Syracuse University to erect the Carrier Dome in 1980 . . . a structure that had no air conditioning. In response to the

The Houston Astrodome, "The Eighth Wonder of the World." The air conditioning obviously worked well; spectators could buy not only cold beer but also hot toddies. (Photo: Jet Lowe, Library of Congress, Historic American Buildings Survey)

Things went even further in 1968 when the Astrodome was joined by the huge amusement park Astroworld. Taking into account the Texas heat, it treated visitors not only to the Alpine Sled, which propelled them through a snowstorm that provided real snow, and the Lost World, which took them on an African river voyage in open-sided boats that provided cooled comfort—but outdoor waiting areas and picnicking sites were air-conditioned, along with "air relief" stations spotted throughout the park. Fred Hofheinz, Houston's mayor, later said that "without air-conditioning, Houston would not have been built at all. It just wouldn't exist, that's all."

However, there was a problem on the horizon. All over the world, but particularly in America, the electric grid wasn't prepared to handle such

inevitable snickers, it was explained that the Dome was used only during the academic year.

demand. Ever since the postwar years, utility companies had eagerly anticipated a growth in power usage, spurred on by a new and tempting variety of electric appliances—including air conditioning. The Edison Electric Institute and General Electric had even turned consumption into its own reward in 1957 when a clutch of media celebrities headed by TV spokesman Ronald Reagan introduced the Gold Medallion Home. It was simple to qualify for this accolade: A homeowner needed only to install enough extra wiring and appliances to use electricity for *every* household function, from cooking to dishwashing to heating and cooling. For his pains, a Gold Medallion Home owner would get a small bronze plaque mounted on the outside of his home, assuring neighbors that the inmates were going to LIVE BETTER ELECTRICALLY. In return, utility companies delightedly estimated that *they* would get an additional $300 per year in electric billings from the homeowner.

But once customers had become used to the idea of "electric living," the power supply wasn't able to keep up with the exploding market. The 1960s saw a whole series of blackouts and brownouts that affected American cities—many of them in the Northeast (and particularly New York City), most of them in the summer months, nearly all of them blamed on people using cooling equipment: "The growing use of air conditioners was reported also to be contributing to the higher demands." "The police blamed widespread use of air-conditioners for the blackout." After highly publicized blackouts in 1959, 1960, and 1961, New York's Consolidated Edison gave a press conference in late June 1963, stating that the chances of another major blackout were "minuscule."

Less than a month later, there was another blackout.

The situation was only going to deteriorate with time, as was demonstrated in November 1965 when a gigantic power failure resulted in a blackout that stretched from New Jersey to Ontario and left 30 million people in the dark. With that debacle still a sour memory, there was worse to come the following July when 100 degree temperatures stretching across much of America caused blackouts and brownouts in a number of cities, including Omaha, St. Louis, Boston, and New York.

The *New York Times* commented, "The heavy drain of air-conditioners was said to be a major cause of these local failures."

Washington's Hot Air (Part VIII)

By this time, so many Americans had air conditioning that there was no longer any real titillation factor offered by knowing that it was used by Washington politicians. A number of those politicos made up for this, however, by treating air conditioning with what could only be called eccentricity.

As the 1960 presidential contest between John F. Kennedy and Richard Nixon was enlivened by a series of four televised debates, the debates themselves were enlivened by reports of the candidates' clashing opinions about the most pleasant temperature for debating. Their reactions to the hot lights were very different, a fact that showed plainly during their first encounter; Kennedy appeared comfortably relaxed, while viewers saw a copiously perspiring Nixon, mopping off in view of television cameras when he thought no one noticed.

A week later, the second debate was held in a studio that had been cooled to 64 degrees. Nixon might have been happy about it, but Kennedy wasn't; a Kennedy aide was dispatched to the basement to find a Nixon aide standing guard over a dialed-down thermostat, which resulted in the Kennedy aide's threatening to bring in the police. The Nixon aide retreated, and by air time the temperature had climbed back up to 70 degrees. The third debate found the candidates on opposite coasts and debating by split-screen: a good thing for Nixon, as it gave him a Los Angeles studio all to himself and he was able to have it cooled to the temperature he found ideal, 60 degrees. He perspired anyway. Kennedy won the election.

When the Kennedy family moved into the White House, they discovered that the Truman-era central system "emitted blasts of either frigid or torrid air." Worse, it seemed to fold under crowd pressure. When the President gave a televised address to the nation in July 1961 from the Oval Office, part of the nationwide reaction was surprise at his uncharacteristically sweaty look. It had been caused by the TV lighting, along with the body heat of nearly fifty people who were crammed into the room behind the cameras, overwhelming the cooling. After that night, Kennedy banned extra guests from White House telecasts and insisted on having a fan at floor level to blow up at his face.

While Kennedy was a Massachusetts native who seemed relatively comfortable with the Washington climate, Lyndon B. Johnson was a Texan who did everything possible to stave it off no matter where he was. His hometown retreat, the "Texas White House," was air-conditioned, and so was its airplane hangar. Even the nearby Johnson City Christian Church (the President's own congregation) received a $4,000 gift from LBJ, specifically earmarked for the installation of cooling equipment. In nearby Austin, presidential offices in the Federal Building were thoroughly air conditioned; as Johnson was told at move-in time, "You'll be able to freeze oranges on your desk, Mr. President." And in Washington, it became common knowledge that LBJ had a clear set of priorities. He would be seen walking around the Executive Mansion, turning off unneeded lights—but at the same time, extra cooling power had been added to the White House system as soon as the Johnson family had moved in. The First Family's quarters were kept chilly enough that the President was rumored to sleep under an electric blanket in summer.

An even greater fan of air conditioning was Johnson's successor, Richard Nixon. As his own feelings about heat had already been reported in the media years before, no one was surprised when Mrs. Nixon showed a reporter his newly renovated White House "hideaway" in 1969 and said, "He has a fire in here every night and plays classical music on the phonograph. We always have a fire, even in summer. Air-conditioning, you know."

At an earlier time Nixon's addiction to chilled air, like Johnson's, might have been seen as a charming quirk. But when Nixon entered the presidency, air conditioning was already commonplace, and the big news angle wasn't its oddity. Utilities were beginning to raise prices, brownouts were becoming an unhappy summertime norm, the Gold Medallion Home had turned into a power-gobbling albatross, and the 1950s prediction of too-cheap-to-be-metered electricity was recognized as an impossible dream. The political viewpoint on air conditioning quickly pivoted away from snickers at its fat-cat luxury to a much more practical angle: worries about its rising operating cost. And there were those who pointed at it as a drain on the country's resources.

With worries about the national energy supply first being voiced to the public, the words "environmentalism" and "ecology" began to pop up, even though they were used primarily as cocktail-party chatter. The first Earth Day took place in April 1970; the U.S. Environmental Protection Agency began its work only four months later with a strong tilt toward energy conservation. And when it came to energy supply, air conditioners, famous for their big electrical appetites, were the first appliances to be examined. In 1972 the Association of Home Appliance Manufacturers announced the creation of the Energy Efficiency Ratio (EER), a "voluntary" rating that would be applied specifically to air conditioners. Measuring energy drain against cooling power, units were rated on a scale from 1 to 10 (that initial batch of air conditioners justified the concern, as not one got a rating better than 8.6). That spring, New York City stores displayed the first EER ratings on air conditioner hang-tags. They confused customers and annoyed salesmen, some of whom were caught impatiently telling buyers that the ratings "weren't important." Even so, the idea spread across the country, and it was so successful that it was soon applied to other home appliances.

Nixon tried to do his bit during the summer of 1973 by making an Energy Policy Statement that announced some conservation measures, Washington-style. One of them was "reduction in the level of air-conditioning of all Federal office buildings throughout the summer." He also recommended to the American homeowner, "Raising the thermostat of an air conditioner by just 4 degrees, for instance, will result in a saving of an estimated 15–20 percent in its use of electricity." Considering his own extensively publicized habit of air-conditioned roaring fires, however, that particular recommendation drew little more than jeers from the public.*

That October, the situation deteriorated when Middle East oil producers abruptly decided on an embargo. With oil shipments seriously curtailed and shortages at gas stations, the United States faced a full-blown energy crisis. The electronic and print media responded with great relish. And air conditioning suddenly became a handy bogeyman for the country's wastefulness; one report pointed out that the nation used only 4 percent of its energy for air conditioning, but American air conditioning represented more than half of all the air conditioning used *in the world*. Everyone lined up to say something about it. Consumer advocate Ralph Nader railed against home units for their "inefficiency." So did *Consumer Reports*. The Federal Energy Office recommended that home thermostats be set to 78 degrees. Authorities ranging from politicians to popular writers lined up for interviews, many suggesting that air conditioning be seriously curtailed or even eliminated. As some of them were nonchalantly suggesting that it be curtailed even in sealed buildings—a stupid suggestion, as occupants would have been not only overheated but also

*This President had another connection with air conditioning, more tenuous but more fateful. During Nixon's 1972 reelection campaign, surrogate G. Gordon Liddy proposed that the GOP stack the deck with a bit of guerrilla warfare. He came up with "Operation Gemstone," a series of acts (including wiretapping, blackmail, and kidnapping) designed to discredit Democrats and their supporters. One component would be "Turquoise"—described as a "commando raid to destroy the air conditioning" at the 1972 Democratic Convention in Miami Beach. "Turquoise" was ruled out as being a trifle far-fetched. The White House settled for something they felt would be more practical: a burglary at Democratic National Committee headquarters, the discovery of which kicked off the entire Watergate probe and ended in Nixon's resignation.

asphyxiated—and as virtually none of these authorities went on the record as stating that they were going to curtail their own air conditioning use, the public gave those ideas short shrift.

The oil embargo ended by March 1974, with a few lasting effects. At one end of the nonsense spectrum, the Federal Energy Office suggested that offices save money with higher summer temperatures and businessmen deal with the extra warmth by going without neckties; this resulted in a grim delegation from the Men's Tie Foundation's immediately being dispatched to Washington for an urgent meeting, after which the Energy Office sheepishly reversed itself and *re*-recommended that gents just go ahead and loosen their collars. More seriously, electric rates went sharply up and didn't come back down. And all the negative attention finally resulted in a new crop of air conditioners that were indeed more energy-efficient, some of them using half the power of their predecessors.

Nixon's replacement, Gerald Ford, continued the message by asking the nation for a 5 percent reduction in energy usage, part of which could come from less air conditioning. This kicked off a flurry of press coverage, including a National Science Foundation survey that asked 602 households for their reactions. It found that more than half of them "saw no connection between the energy shortage and their own households," only 7.6 percent of them knew that a higher EER meant a more efficient air conditioner, and 36 of the households felt there was really no energy crisis at all.

Americans got a truly conservation-minded president in Jimmy Carter, even if they didn't quite know what to make of him. As one story went, he had an early run-in with home comfort technology when he was Governor of Georgia and decided to conserve energy by turning down the executive mansion's wintertime thermostat. In response, heating bills went up instead of down. Discovering that the heating system was electronically linked to its air conditioning, would self-adjust if it were tampered with by the human element, and that it would cost more to remove the system than it would to pay the higher utility bills, he gave up on the idea.

Once Carter assumed the presidency, though, he was determined to prove the benefits not only of conservation but also of alternative energy sources. At his January 1977 inauguration, the nation learned that baseboard heaters in the presidential reviewing stand were powered by solar energy. The very next day, he gave a statement in which he recommended that home thermostats be turned down to 65 degrees to save fuel, announced that this would be mandatory in federal buildings, and said,

"Today's crisis is a painful reminder that our energy problems are real and cannot be ignored." Less than two weeks later he repeated this in a televised address to the nation, emphasizing the idea of pitching in by wearing a cardigan sweater.

But the 1970s had been dubbed the Me Decade, disco fever was in, and selflessness was distinctly out. Carter's address resulted in a short-lived fad for cardigan sweaters among the executive set, but otherwise not many people cared about his message, or for that matter even trusted it. A *New York Times*/CBS poll taken in late 1977 "shows that 51 percent of the public does not believe Mr. Carter's assertions that there is a real shortage." There were few other visible reactions to the news, except for a sudden burst of interest in ceiling fans; they might have been old-fashioned technology, but they used a fraction of the electricity of air conditioning. There was even a *Wall Street Journal* interview with a dealer who was capitalizing on the upscale market for refurbished antique speci-mens. "They're novel, they're efficient, and where can you find an air conditioner with the character of a fan?"

Undaunted, the President tried other ways to set an example. Six wood stoves appeared in the White House and the presidential retreat Camp David; Carter even had $28,000 worth of solar panels installed on the White House roof to provide energy for water heating, showing them off in an *al fresco* news conference. Not only did they not serve as a model for other installations, but they became punch-lines for late-night comics and the Republican opposition. One right-wing commentator sug-gested that Carter campaign literature would make perfect firestarters for the wood stoves.

Up until now, Carter's sights had been trained on heating, but air conditioning got involved in May 1979 when he pushed through an initia-tive to "order public and commercial building thermostats set no lower than 80 degrees in summer." The public revolted, as did two federal judges in Albuquerque and Beaumont, Texas, who had their courtrooms set to 74 degrees and 70 degrees, respectively, and made sure that every-one in the country knew it. By the end of June, the initiative was "modi-fied" to allow buildings to be cooled to 78 degrees. Even so, it went unenforced and basically unnoticed.

Carter continued to have bad luck, much of it centered on energy policy. When a repeat oil embargo hit the nation in 1979, it was received with impotent anger by citizens who didn't want to hear that they had been warned about it. A 1980 heat wave lasted for most of the summer, centering on the Midwest and the South and breaking innumerable

President Jimmy Carter, demonstrating the new solar panels to reporters. At the time, he said that the energy crisis required "the moral equivalent of war." Critics snickered and responded with the acronym MEOW. (Courtesy: Jimmy Carter Presidential Library)

records; Dallas alone experienced 42 consecutive days of 100-plus–degrees temperatures. The public was helplessly transfixed by the heat, the inevitable brownouts, and the $20 billion loss from destroyed crops and livestock. Worst of all was the rising death toll, which would eventually reach more than 1,200. There was bitterness, some of it unreasonably directed toward Carter as a doomsayer whose prediction had happened to come true.

In the end, air conditioning got to Carter in another way. Throughout the 1970s and before, the country had been experiencing a major population shift, with millions of people moving to the "Sunbelt" states of the South and Southwest. Scientists, sociologists, even pop historians agreed that one of the major reasons for the migration was the rise of air conditioning, which had made those states attractively livable—and attractive particularly to older, more conservative voters. This population shift redrew the political map of the nation, making it no surprise that former Gold Medallion TV spokesman Ronald Reagan handily trounced Carter in the 1980 election.

An allergy sufferer, Reagan was an enthusiastic fan of air conditioning. A good thing, too; by the summer of 1986, the White House system was being completely overhauled once again, having been deemed "inadequate" by the National Park Service.

As to Carter's solar panels, they had been languishing on the roof since Reagan's election. During the overhaul, they were unceremoniously taken down.

Problems . . .

Anyone who had thought during those years that comfort cooling would somehow disappear for the good of humankind was mistaken; the 1980 census showed that 57 percent of American homes had air conditioning, a significant jump from previous indicators. The energy crisis and its belt-tightening created a lull in the industry, but it vanished as the world economy rebounded during the bonanza of the following decade, a period so spectacularly excessive that it was nicknamed the Roaring Eighties. But even though HVAC (for Heating/Ventilating/Air Conditioning) had been around for three-quarters of a century and was an absolute necessity for most buildings, at the same time it was receiving more criticism, including outright calls for its abandonment, than ever before. And not for its cost.

At the beginning of the 1970s a few lone researchers had been shocked to notice changes in the Earth's upper atmosphere, in particular a thinning of its ozone layer. At first these changes were thought to be the result of exhaust vapor coming from supersonic planes. Then the magazine *Nature* suggested that methane was to blame; one expert claimed that "the flatulence of domesticated cattle is a major contributor." Threatening as gassy livestock might be, it turned out that chlorofluoro-carbons, CFCs, were an even greater culprit. And one of the most widely used CFCs was that "outstanding scientific achievement" of 1928, "Refrigerant R-12," alias Freon. Ever since Thomas Midgley had first wowed the audience at the American Chemical Society with his blow-out-a-candle-with-the-gas trick, Freon had virtually locked up the market not only as an aerosol propellant but as the go-to refrigerant for air conditioners everywhere. However, that good fortune had a nasty side effect. Sprayed from aerosol cans, leaking from auto air conditioners, and escaping from damaged or junked cooling coils, Freon was drifting straight up into the stratosphere. There it would stay, maddeningly stable for up to a century, ever so slowly degrading into chlorine . . . and, ultimately, eating away at ozone. One report estimated that the atmosphere was playing host to 2.5 billion pounds of Freon.

The story hit the mainstream press in late 1974, electrifying the scientific community. As to consumers, they didn't know what to think. The initial focus was on all the Freon being sprayed directly into the air from millions of spray cans (the *Minneapolis Star* jolted its readers to attention with the headline "Can Dry Armpits Mean World Crisis?"); air conditioners were assumed to be less of a problem, as the Freon sealed inside cooling coils was meant to stay put.

By 1978, manufacturers had removed Freon from aerosol cans. But the air conditioning industry was forced to move more slowly. With roughly 60 percent of Americans using some form of air conditioning in their homes, and all of them using Freon, an outright ban would put every one of those units under a death sentence. A new crop of refrigerants was developed, but their own chemical idiosyncrasies meant that they couldn't be injected into existing systems. Compromise was the key. Truth was, Freon wouldn't ever be able to disappear completely from the marketplace.

From the early 1990s on, air conditioners of all types used other refrigerants, but Freon had to be available to service older machinery. Its production was slated Absolutely to End in 1994, a deadline that was then stretched to 1995. And as it couldn't be released into the atmosphere, from then on it would be extracted from defunct units, as cautiously as any biohazard, and recycled for reuse. Because of the fuss, its price-per-pound shot up from $1.00 to $15.00. Many auto air conditioning shops (which dealt with three-quarters of the Freon being used) stockpiled as much as they could handle—not an easy task, as the cans tended to leak—and there was even a black market for bootleg Freon. In 1995, the *San Francisco Chronicle* interviewed a U.S. Customs official who said, "Freon is right behind drugs as the fastest growing area of crime."

At the same time the Freon story was ramping up, there was another, more ominous problem. In the summer of 1976, Philadelphia hosted an American Legion convention at the grand Bellevue-Stratford Hotel: a spectacular turn-of-the-century architectural confection, a favorite of Main Line society, and, since the 1960s, completely air-conditioned. Starting two days after the convention began, some Legionnaires developed flu-like symptoms that escalated to full-blown pneumonia. Within a few weeks, more than 200 of those people would be sick enough to need medical treatment, and 34 of them would die. In the panic to find a cause, theories abounded that the Legionnaires had been felled by pesticides, biowarfare, nickel poisoning, parrot fever, "Refrigerant F-11," and even "the fatal disease pantosomatitis" (which turned out not to exist).

After nearly six months of morbid press coverage and frantic search, the answer was found: *Legionella pneumophilia*, a previously undiscovered organism that was breeding in the cooling water of the hotel's air conditioning system. Guests, employees, and even passersby on the street were routinely exposed to the conditioned air, which meant that they were inhaling microscopic droplets of the water along with a dose of *Legionella*.

Air conditioning installations the world over were quickly tested for bacteria, with frightening results; it turned out that 40 to 60 percent of them harbored *Legionella*. Worse, investigators dug up the information that there had been other cases of epidemic pneumonia starting in air-conditioned buildings, cases that went back to the 1950s.

People might not have given much thought to the fastidiousness of air conditioning machinery in days past, but they did now. Especially when they read interviews with public health experts that disgusted them: "After 10 or 20 years, almost all these systems have an accumulation of slime, fungi or bacteria . . . a prescription for trouble." Another consultant told of inspecting ducts that were completely choked with years of dust, along with an occasional discarded lunch bag or beer can: "Ductwork is out of sight, and usually built without simple ways to gain access, so people just ignore it." Aggravating the problem was the fact that, in the face of the energy crisis, building managers were trying to save money by bringing in less fresh air (which had to be heated/cooled) and recycling more of the inside air—a process that concentrated whatever was contaminating the air in the first place. By the early 1980s, a specific term was being used to describe the problems that came with a building whose air was coming in too mechanically: Sick Building Syndrome.

In 1986, *Consumers Union* named air conditioning as one of the top fifty inventions "that have most influenced our lives." Even so, it was becoming obvious that the decades-old fantasy of a tightly sealed building, improving on the outside air by replacing every breath with "manufactured weather," wasn't practical. Or healthy.

Washington's Hot Air (Part IX, Green Edition)

Despite the crackle of bad press, air conditioning was still viewed with applause, as well as an occasional spark of surprise at the tricks it could pull. When the huger-than-huge 8,000-seat Basilica of Our Lady of Peace of Yamoussoukro was being built in the Ivory Coast in 1989, the *New York Times* reported, "To save electricity, engineers will cool only a 15-foot-high cushion of air, leaving the 380-foot-high dome to warm to tropical levels." In 1993, after the Sistine Chapel had its Michelangelo frescoes restored, they were protected by air conditioning that provided some of the most sophisticated climate control outside of a laboratory; Pope John Paul II declared that the engineers "have become co-workers with the painters in preserving these creations." The following year, the Britain-to-France railroad tunnel that traversed the English Channel—the Chunnel—was air-conditioned. Not merely the trains: the tunnel itself.

There was a persistent drumbeat of media information on all these topics, and the George H.W. Bush administration wasn't immune; this was the President who called for the recycling and eventual phase-out of Freon, and for that matter Bush had the White House cooling system "renovated" in 1991. But it was his successor, Bill Clinton—with an assist from Vice President Al Gore, an aggressive environmentalist—who made more of an impact. On Earth Day 1993 Clinton announced the "Greening of the White House," a project to shape up the Executive Mansion's environmental performance. The project saved $300,000 each year in household-running costs, and part of it was yet another retrofitting of the air conditioning system that aimed for 20 percent more efficiency. (From a personal standpoint, this was a good idea. Even when Clinton was a newly elected President, the *Washington Post* reported that he was allergic to a host of irritants, among them ragweed, pollen, and the family cat, Socks—and the paper helpfully suggested that he use air conditioning as much as possible.) But ordinary citizens who didn't have high-level government access weren't left out; in 1995 the Energy Star rating, originally designed to reward the most energy-efficient electronic equipment, was expanded to take in air conditioners. And the EER had been not only improved but also expanded into the SEER, the Seasonal Energy Efficiency Ratio.

True to Washington form, those ratings became a political football. Three days before the end of Clinton's term, the Energy Department announced a rule to begin in 2006 that would raise the efficiency of central systems by 30 percent. Less than four months later, the George W. Bush administration scaled back the efficiency mark to 20 percent.

This wasn't to say that Bush was completely averse to environmental causes. When his home was built in Crawford, Texas, its heat and air conditioning came from a "geothermal" installation that used the laws of physics, aided by a web of underground piping, to extract heat from a building and move it into the ground; he called the system "environmentally hip." As well, during his administration 167 solar panels were installed on the White House roof, generating power for an outbuilding and heating his swimming pool.

Barack Obama entered the White House in 2009 with an avowed commitment to greening; an idea, and a word, that finally had become chic. He pushed the idea of conservation via green home remodeling—telling an audience, "Insulation is sexy stuff"—and rebates for energy-conscious remodeling (the program was nicknamed "Cash for Caulkers"). The White House joined in the updating, and more solar panels appeared on the roof.

It was generous of the President to take an interest, as air conditioning evidently wasn't important to him. The *Daily Telegraph* reported that he was "known to keep his Oval Office at temperatures reminiscent of his native Hawaii." In fact, his personal aide Reggie Love told ABC-TV, "The thing that used to kill me is that the guy loves to ride around with the AC off in the summertime. And I get hot. And I'm sweating. And I'm like, it's 80 degrees in this car. I'm going to pass out."

. . . and Possible Solutions

Around 1979, American businessman/traveler Christopher Hyland was organizing a lengthy vacation that would include some railroad travel through India. Discovering that the Maharaja of Jodhpur owned a spectacularly outfitted parlor car, he asked to charter it for his party. Not only was the Maharaja willing, but he threw in a stay at his palace.

When the Hyland party arrived, they discovered that the palace had its own way of keeping cool—a *tatty* for each guest room. In a concession to the twentieth century, there were no servants stationed to wet the *tatties*; they were automatically sprinkled with water that came from an arrangement of pipes running above each window.

After a surprisingly comfortable night's sleep, they were told, "Don't worry—the train will be very cool. It has air conditioning." Sort of. Each sleeping compartment was equipped with a large porcelain vat, filled with chunks of ice, and an electric fan blowing on it.

"I got into bed," recalled Hyland, "looked at that fan, and thought, 'This could be 1905.'"

While the Maharaja's household was simply doing for its guests what it had always done, in another way this was significant. Throughout India, and everywhere else in the world, mechanical cooling had been growing in leaps and bounds; but at the same, there was a growing feeling that standard air conditioning was no longer the best way of dealing with heat. In the effort to find an alternative system, perhaps older methods, or extremely uncommon ones, would be worth another look.

As with earlier years, when Carrier's Apparatus for Treating Air had to share the limelight with the Modine Ice-Fan and the Nevo, by the turn of the twenty-first century there had already been a string of other inventors with ideas of their own for cold-making. Some of those ideas were so cutting-edge that they couldn't make it into the average household. David Sarnoff's 1950s-era prediction of "noiseless" air conditioning "with no moving parts"—"thermoelectric cooling," using only the movement of electrical current to carry away heat—turned out to devour even more electricity than standard air conditioning; in the consumer market it never could get any further than the novelty stage, a single-tray ice cube maker that was installed in a few Chicago hotel rooms. But if it was too expensive to consider in a large form, it would prove invaluable in miniature. Once computers became universal, tiny thermoelectric air conditioners proved to be just the thing to cool their entrails.

Even more ahead of the curve was the idea of "thermoacoustic" refrigeration, which used high-intensity sound waves to "drive pressure changes that result in cooling." Discovered in the nineteenth century, it became extraordinarily popular in the 1990s as a research subject, studied at universities from Canada to Australia. While the verdict at first was that "experimental thermoacoustic systems are not nearly as efficient as conventional air conditioners," in 2004 the technology had come far enough that a thermoacoustic freezer was demonstrated in Ben and Jerry's ice cream shops all over New York City. And *Appliance Design* in 2008 mentioned tests of "a commercial rooftop A/C unit."

With literally hundreds of refrigerants showing up to replace Freon, it was only a matter of time before their own drawbacks would become apparent; there was the much-publicized example of "Refrigerant HFC-134a," the standard replacement for Freon in automobile air conditioners, which turned out to be kind to the ozone layer but less than kind when it came to contributing to the "greenhouse effect" and global warming. When it became known that the same could be said for HFC-410a, R-125, R-32, R-143a, and a whole host of their cousins, some old-fashioned compounds were recommended as alternatives. Carbon dioxide hadn't been a serious contender for decades, but it received another look, if only because it had *no* potential for harming the ozone layer. There was still more nostalgia when the International Institute of Ammonia Refrigeration, which had been marking time since the early 1970s, bounded to the forefront with publications such as "Ammonia: The Natural Refrigerant of Choice," which made a virtue out of everything that everyone had ever hated about ammonia, including its knock-'em-dead pungency: "[I]t has an ozone depletion potential of zero. . . . Ammonia refrigeration has a proven safety record, in part because of the physical properties of ammonia, not the least of which is ammonia's well-recognizable and easily-detectable odor. . . ."

Then there was the substance that had been classified as "Refrigerant R-718"—water. Even though the thoroughgoing chill of standard air conditioning had long since invaded the Southwest, evaporative (swamp) cooling never completely vanished from the scene. With its attractively low operating cost, during the last decades of the twentieth century it began to make something of a comeback and even to spread to other continents. Manufacturers from the United States to China offered coolers in every variety, including small-scale portables, large portables for outdoor use (one Indian company recommended its portables "for wedding parties"), even gigantic units that could cool entire factories. But while they were cheap to buy and cheap to run, they were still at the mercy of outdoor humidity; not-dry-enough meant not-cool-enough.

That problem was mitigated when several companies played around with the evaporative cooling concept, formulating ways to chill the air while wringing moisture from it at the same time. One of the most promising efforts, sponsored by the U.S. government's National Renewable Energy Laboratory, made news in the summer of 2010 with the unveiling of the Dessicant Enhanced eVaporative air conditioner, or DEVap. Rather than cooling coils or refrigerant, the DEVap relies on a special polymer membrane, water, a desiccant solution to absorb moisture, and two separate streams of air to produce its *coolth*. The best news about this extraordinary system—at this writing, still in the testing stage and three to five years away from production—was that it will be able to cool and dry air while using up to 90 percent less energy than the most efficient standard air conditioner.

But there were systems that could, in theory, use little or no electric energy at all. Solar energy had been researched for decades, at venues ranging from the University of Florida to the government of Saudi Arabia, and some of this work had borne fruit; the United States Pavilion at the 1982 Knoxville World's Fair was outfitted with 5,000 square feet of solar collectors to power the building's air conditioning and hot water

The heart of a DEVap. *Left*, a closeup view of the "membrane based heat mass exchanger"; *right*, a prototype DEVap being laboratory-tested for effectiveness and energy consumption. (Photo: Dennis Schroeder/NREL)

systems. Even more intriguing were other solar-based systems that substituted a "thermal compressor," which operated without any of the usual refrigerants, for the electrically driven model.

Many of these ideas were prototypes or untried concepts, and without serious financial backing they remained for years little more than curiosities. Things changed, though, when the concept of "green architecture" moved from discussion into reality. In 1992, the National Audubon Society moved its New York headquarters to an abandoned 1891 building on Lower Broadway. As the Society had for several decades been concentrating on the country's ecological situation, the building would set an example with environmentally conscious remodeling: low-toxicity paints, reclaimed woods, and subflooring made from recycled newspapers. But the job went further with features that not only saved energy but also boosted the indoor comfort quotient: heat-blocking windows, ventilation that completely changed the indoor air six times more often than city regulations demanded—and, as a crowning touch, the *Christian Science Monitor* reported "a natural-gas, heat-transfer system for its air-conditioning system, instead of Freon." The public was fascinated and the real estate community even more so, as the renovation was projected to save $100,000 each year in energy costs. Within a year, it was pronounced "one of the lowest energy-consuming office buildings in the country."*

By the time the World Congress of Architects met in Chicago in mid-1993, 5,000 attendees discussed the situation thoroughly and came to an uneasy consensus; "glass energy hogs" around the globe might be stunning design statements, but they were economic as well as ecological disasters. If anyone needed persuasion, there was a perfect example within walking distance: Chicago's own State of Illinois Center. The circular 17-story steel-and-glass colossus had been designed in 1985 with double-glazed insulating windows that were vetoed—ignoring the lessons of Le Corbusier—because of budget problems. Instead, the building was constructed with single glazing, which meant that tenants had nothing other than the air conditioning to shield them from Chicago's summer heat. It didn't help. Even with the system running full blast, indoor temperatures occasionally reached 95 degrees. Management added $10 million in extra cooling machinery and installed 1,671 sets of Venetian blinds, but tenants still referred to parts of the building as "Death Valley" and "The Tropical Zone."

With more than 40 percent of the energy consumption of some buildings going for air conditioning, developers and architects realized that some crucial changes had to be made. If for no other reason than the savings potential, the business community fell into line with this way of thinking.

A particularly imaginative solution to the problem was erected in 1996: Eastgate Centre, a combination office block and shopping mall in Harare, Zimbabwe. Even though the city has its own share of glass energy hogs, architect Mick Pearce moved resolutely away from that ideal in every aspect of the project from design to construction. Significantly, he considered Harare's high elevation, which gave it comparatively temperate summers along with cool nights, and concluded that a building designed with the help of twentieth-century computer models could take advantage of some nineteenth-century climate control ideas, in particular thermo-ventilation. Eastgate Centre consists of two side-by-side buildings

---

*By happy coincidence, the building had been designed by George B. Post, the selfsame architect who had the New York Stock Exchange to his credit. As Post had been there for one of history's very first large-scale air conditioning jobs, it was fitting that his name was now connected to a project that tried to modify the process to suit modern society's new ecological needs.

The State of Illinois Center—or, as one writer dubbed it, "a vast greenhouse filled with overheated bureaucrats." (Photo: Primeromundo)

with a breeze-gathering space between them; thick concrete walls, which deeply shade windows and avoid direct glare; and an extraordinarily sophisticated "natural cooling" system that operates twenty-four hours a day to shunt the air upward through each floor and out through a series of chimneys. The system is so carefully planned that it keeps the complex at a uniform temperature while using only 10 percent of the energy spent in standard air conditioning.

These ideas worked well enough that they were adapted for use in London's 2001 Portcullis House, which was built as additional office space for Members of Parliament as an across-the-road neighbor to the Palace of Westminster—and appropriately enough, in sight of its Central Tower, that nineteenth-century monument to thermo-ventilation. Indeed, Portcullis House has no conventional air conditioning; it relies instead on a combination of sun-shading construction, triple-glazed insulating windows, fourteen chimney-like ventilation stacks, and a system of fans. Fresh air is drawn in at the roof, routed to the lowest level of the building, cooled if necessary by being blown over pipes containing water pumped up from deep wells, then directed up through the building before exiting through the stacks. In practice this system has functioned efficiently in some ways but not in others, with uncomfortable humidity levels showing up as a recurring problem. And the building surprised its admirers when it was placed on a "Dirty Dozen" list of distinguished London buildings that failed to hold their heat during the cold months. Still, it was viewed as a possible step in the right direction.

In the meantime, the steel-and-glass building proved that it could use its resources more efficiently when Frankfurt's Commerzbank Tower opened in 1997. This "ecological skyscraper" was a remarkably sensitive one in terms of its energy use. It even had windows that could open.

"The tallest naturally ventilated building in the world" was designed to take full advantage of the outside atmosphere. For maximum airflow, it was designed with an atrium running up its entire fifty-six stories. And in a surprising reversal of usual procedures, its windows worked. Of course it couldn't be a complete return to the old days; to coordinate the times at which air conditioning would and wouldn't be operating, and

An aerial view of Portcullis House, highlighting its ventilation stacks.
(Author's collection)

windows could and couldn't be opened, each room had its own "control panel" (red light, please don't touch; green light, go right ahead). As to its cooling, the Tower's brochure described the arrangement: "The building is cooled by a water-filled chilled ceiling system instead of the normal, problematic air-conditioning system. . . . Employees are able to control the temperature in their office[s] individually within a given range. . . . On roughly three-quarters of the days in a year, employees can regulate ventilation themselves by opening or closing the windows individually."

Even though Frankfurt's moderate summers didn't make for the most rigorous test of a building's comfort capabilities, architects and critics from around the globe visited Commerzbank to see how well the Tower worked. While the *Chicago Tribune* wasn't bowled over, either by the architecture or by the temperature control (and it snickered that the designers had perhaps gone overboard on the ecology, as the restroom washbasins supplied only cold water), it admitted that the Tower "demonstrates, as only a major monument built by a major corporation can, that the fledgling movement known as Green Architecture has moved decisively into the mainstream."

Before the Commerzbank Tower had even opened, it was being one-upped in New York by Four Times Square. When it opened in 1999, neither its 48-story height nor its steel-and-glass modernism was out of the ordinary; but as "the world's first green skyscraper," it was an absolute sensation. The exterior was clad in high-performance glazing that "let in light without affecting inside air temperature"; the top 19 stories contained solar panels to generate electricity; and when it came to air comfort, Four Times Square, which would have to contend with far hotter summer weather than any building in Frankfurt, went the distance. The building provided tenants 50 percent more fresh air than city codes required, super-filtered and taken from 700 feet above the street. And cooling came from "gas-fired absorption chillers," no CFCs needed.

With the construction of Four Times Square and the publicity that surrounded it, and then the worldwide introduction in 2000 of LEED Certification (for Leadership in Energy and Environmental Design) to reward energy-conscious construction, it became fashionable—hence urgent—for a skyscraper, and particularly its climate control, to be green. In the first decade of the twenty-first century, ecologically friendly buildings arose everywhere, some of them gigantic, many with ingenious air conditioning schemes. "The London Gherkin," 30 St Mary Axe, might have been whimsically shaped like a glass pickle, but its cooling plan was absolutely serious, with windows that opened automatically to "augment the air conditioning system with natural ventilation, an occurrence anticipated to save energy for up to 40% of the year."* In the African desert–dry climate of Windhoek, Namibia, developers of the 21-story Mutual Tower realized that evaporative cooling would not only cost a fraction of standard air conditioning but would also make for healthier air, one of the touches that earned it the nickname "Namibia's Tallest, Greenest Building." And in the Middle East, where 70 percent of the region's energy usage went for air conditioning, the Bahrain World Trade Center worked into its design three 95-foot-diameter wind turbines that could generate 15 percent of its electricity; better still, it tapped into Bahrain's "district cooling" system, which took advantage of the island locale by using seawater as the source of cold that provided air conditioning to a whole series of buildings. Even super-skyscrapers became green; in 2001, Taiwan's 101-story Taipei 101, for a few years the world's tallest building, installed an immense plant that manufactured ice during overnight hours, when electricity was cheapest, and used the ice throughout the day as a cold source that allowed the building to use a fraction of the air conditioning equipment it would otherwise have needed. This gambit was popular in other places; buildings from San Francisco to New York to Barcelona have updated their systems and slashed their energy usage with stored-ice cooling.

Perhaps the ultimate example of corporate cooling with an environmental twist would be found outside of Abu Dhabi—Masdar City, the planned community that was announced in 2006 and had its first buildings ready for occupancy three years later. Its six square kilometers were intended as the last word in green construction and green living: sustainable building materials, electricity from solar panels, and cars replaced by battery-powered "Personal Rapid Transit" vehicles. And to cope with the desert heat there was a mix of ancient and ultra-modern strategies: High fencing around the city's perimeter (to block hot breezes), narrow streets (to provide maximum shade), and "wind towers" (to shunt cooler air down to street level) were partnered with "solar umbrellas" in the city center that automatically opened during the day and closed at night, solar-powered air conditioning, continuing work on geothermal cooling . . . and

---

*In 2012, this author visited the Gherkin to get a first-hand look at the system and discovered, when an elevator got stuck between floors for an hour, that the system wasn't quite universal throughout the building. During that hour, the occupants learned that the cars had been built without any cooling—or, for that matter, any ventilation at all.

30 St Mary Axe, at a moment when its windows are closed.
(Photo: Aurelien Guichard)

sometimes the ancient and the modern were combined; one of the central wind towers displayed an LED "beacon," actually a tattletale electric sign that glowed red to inform city residents that someone's energy consumption was too high.

As Masdar City headed toward its estimated completion date of 2016, proponents were ecstatic, while at the same time skeptics called it a "walled green Utopia," "the ultimate gated community," and "a green Disneyland." Furthermore, they wondered if it could possibly happen anywhere but in a monarchy with plenty of unused real estate.

The greening of commercial buildings, specifically the greening of their temperature control, was a very encouraging trend. But at the beginning of the 2010s, it hadn't yet shown up in a home version. No doubt the problem is exactly the same one that bothered the air conditioning industry in the 1920s—new technology is money-saving only when it comes in a giant economy size.

Conclusion

By the year 2014, we've learned that more than 87 percent of American homes have air conditioning "in one or more rooms." Some estimates are closer to 90 percent. Cars have an even higher percentage, more than 98 percent.

But if the United States had always been the air conditioning leader, the world had recently begun to catch up with a vengeance. In 2008, *Forbes* reported that an astounding 20,000,000 air conditioners were being sold in China each year; by 2010, *American Scientist* noted that the number had climbed to 50,000,000. At the same time, in the searing desert climate of oil-wealthy Dubai, the superluxury hotel Palazzo Versace raised international eyebrows when it was announced that the hotel's private beach—its *sand*, that is—would be refrigerated by unobtrusively buried cooling pipes. With the downturn in the world economy, the air-conditioned beach was apparently put on hold. But disappointed members of the glitterati could still console themselves with a trip to Ski Dubai, a sprawling indoor facility that offered more than five acres of artificial snow, a chairlift, and "penguin encounters," with the whole setup maintained at a frosty 30 degrees.

On the Indian subcontinent, air conditioning has become not only a ubiquitous comfort feature but also a bona fide status symbol, and without a doubt a snazzy addition to a dowry. One young woman recently advertised in her local paper for a husband; listed among her personal attractions were "a TV set, an air conditioner, a chicken, a goat, and a scooter." This situation won't be changing anytime soon. *Scientific American* noted that twenty-eight out of the thirty largest cities in the world were in tropical climates. And as they develop, "the demand for air conditioning in these gigantic mega-cities . . . is going to skyrocket."

True enough. Worldwide, mechanical cooling is becoming routine. Not only buildings, but also transit systems around the globe are constructed with built-in air conditioning, from Kuala Lumpur and Singapore

When Willis Carrier extolled the wonders of "Manufactured Weather," he
probably hadn't envisioned anything like Ski Dubai.
(Photo: Filipe Fortes/fortes.com)

to Bilbao and San Francisco; even the Copenhagen Metro found it necessary. And significantly, London's system, which pre-dated nearly every
kind of cooling equipment, is now finding it necessary, too. In past times,
the average British summer hadn't produced too much discomfort in the
Tube, and in fact one turn-of-the-twentieth-century advertisement reputedly touted the Underground as "the coolest place in hot weather." In the
twenty-first century, it's not the case. Temperatures in trains and stations
have risen so alarmingly that in recent years the Underground begins
each summer by plastering stations with posters detailing its "Beat the
Heat" campaign: "We know the Tube can be uncomfortable in hot
weather. . . . Always carry a bottle of water with you. . . . We are commit-
ted to finding a solution to keeping the Tube cool during hot weather and
are continually investigating new technologies to achieve this aim." While
some lines have been equipped with standard air conditioning, not all of
them can be treated in the same fashion—a number of London's oldest
subway tunnels are an extraordinarily snug fit, providing little room for
anything other than the trains that travel through them, and definitely
not for the waste heat that would be spewed into them as a byproduct of
air conditioning. Without enough room to dissipate, that kind of heat
could actually damage tunnel walls. Consequently, other methods need to
be considered. Experts are currently studying such ideas as air blown
over cold water–filled pipes to cool the stations, mechanically chilled air
blown down escalator shafts to cool the platforms, even air blown over
the underside of a train as it brakes to cool the wheels and reduce the

heat in each station. This dilemma is puzzling, but not unusual. Other transit systems around the world have achieved comfortably cool trains but at the price of blazing hot stations.

Discomfiting as it might be to hear of London struggling with a heat problem, a more alarming piece of news came from the Canadian far north. Even though the Inuit village of Kuujjuaq is closer to Greenland than it is to Montreal, in 2006 residents noted that their summer temperatures had been steadily breaking records. And the mayor ordered ten air conditioners for his office workers.

There is indeed a gradual rise in temperature all over the globe. It is a frightening fact of modern life; also a source of political controversy, as some authorities call it Global Warming or Climate Change, while others prefer to name it Freak Occurrence. Whatever the label, the unfortunate truth is that heat waves occur now in areas of the world where heat waves rarely occurred before, and with a sinister new kind of ferocity. We hear reports of heat in Bulgaria that has literally melted underground electrical cables, of heat in Great Britain that was intense enough to shatter a glass panel in the atrium of the "naturally air conditioned" Portcullis House and that has softened tarmac roads so dangerously that spreader trucks had to be sent out in midsummer to "grit" the sticky pavement, of the fact that temperatures in Australia have regularly been climbing to such unprecedented levels that the Australian Bureau of Meteorology was forced in 2013 to add an additional color to its heat maps.

There are many possible causes for this trend—for which the steeply increasing demand for machine-made cold is, ironically, a prime suspect. Air conditioning works to absorb the heat from the indoors and shunt it to the outdoors, a simple process of physical science, and one that occurs every moment in the machinery of every air conditioner in the world. Can it be that this process, multiplied a billionfold, is the cause of the rise in the world's temperature? Or could it be the fault of the greenhouse effect? The ozone layer? The first signs of another Ice Age? Politicians and news commentators have plenty of fodder for arguments over whether air conditioning is really to blame for the warming trend.

Unhappy as the news might be, underscoring it for more than a decade has been a series of power outages, not only in the United States but throughout the world. Many of them occurred during the hot months and were blamed on the increasing use of air conditioning. One of the most dramatic occurred in India during the summer of 2012, a two-day "complete collapse" of its electrical grid that left a tenth of the earth's population, more than 600 million people, without power. This was extreme, but not unexpected; the booming Indian market for air conditioning means that there simply isn't enough electricity to go around when demand rises, and blackouts are common happenings. Large buildings on the subcontinent have become used to this. They simply install diesel generators for backup power. Environmentalists glance at the exhaust and wring their hands.

Average people tune out all of that news. And as temperatures continue to break records, they react by turning on the air conditioner.

Some experts tell us that this is a bad habit. For reasons of ecology or economy or even The Weakening of the Species (it's been said that the summertime dependence on cooled air makes *homo sapiens* less able to withstand heat), they see air conditioning as a grievous fault of an invention, one that should disappear from the face of the Earth for the good of the planet.

Really?

Perhaps those experts should reconsider that idea. Other than the fact that a great many of the world's buildings would be completely unventilated, and hence uninhabitable, there's a more disturbing example of what could happen to *people*: that ongoing spate of heat waves, and particularly the death tolls attached to them. The estimates always vary widely, something that is annoying from a statistical viewpoint. And grim, when it becomes clear that those numbers represent people who died because they couldn't withstand the heat and couldn't find a way to escape it—between 5,000 and 10,000 throughout the United States in 1988; 750, all dying in a single week in Chicago in 1995; and, most horribly, there was the heat wave that engulfed Europe in 2003. On the Continent, as compared with the United States, few commercial buildings and even fewer homes had ever been air-conditioned. After a month of scorching glare with no relief, more than 35,000 (some say 70,000) Europeans died.

In the face of such statistics, it doesn't matter whether this is a trend or a meteorological fluke; people need a weapon against heat.

But this may not turn out to be the "air conditioner" as we know it. Willis Carrier's machine was an absolute wonder when it was first unveiled. Still, without some tweaking, it's a technology that seems to have outlived its time. It won't be usable in a world that needs to worry about deteriorating atmosphere, or dwindling power supplies.

But love it or despise it, air conditioning has insinuated itself into the world's day-to-day existence as a business aid, a therapy, and a plain necessity. And it's not going away—quite the opposite. If the original design for "manufactured weather" has turned out not to be practical in the long term, certainly the ongoing demand will spur the development of a usable substitute, even more than one, before long. Whatever that system might be, in all likelihood it will make its first appearances in a large-size commercial form, same as Carrier's. Then, also like Carrier's, it will shrink in size and price for home use. After that, the compressor-powered, refrigerant-cooled air conditioner will be an obsolete curio, same as the Victrola.

As fanciful and Tom Swift–ish as some of those ideas might undoubtedly seem right now, people thought the same of Willis Carrier. And Steele MacKaye. And John Gorrie.

Notes

1. Ice, Air, and Crowd Poison

1. Thomas Hood, "Hot Weather at the Play," from *The Works of Thomas Hood* (London: E. Moxon, Son, & Co., 1873).

2. Lady Duffus Hardy, *Through Cities and Prairie Lands: Sketches of an American Tour* (R. Worthington, 1881).

3. *Charleston Mercury*, August 3, 1822.

4. *New York Daily Times*, July 28, 1856.

5. Anonymous, *Hardships Made Easy* (London: Blackwood, 1861).

6. *Philadelphia Daily Dispatch*, June 15, 1855.

7. *Music and Art*, August 1896.

2. The Wondrous Comfort of Ammonia

1. *New-York Tribune*, September 4, 1881.

2. *New-York Tribune*, July 10, 1881.

3. Ibid.

4. *New-York Tribune*, July 10, 1881.

5. *Ice and Refrigeration*, September 1891.

6. *Scientific American*, July 30, 1892.

7. *Chicago Daily Tribune*, June 25, 1893.

8. *Chicago Times-Herald*, reprinted in *The Colliery Engineer and Metal Miner*, February 1896.

9. Adolf Bernhard Meyer, *Studies of the Museums and Kindred Institutions of New York City, Albany, Buffalo and Chicago, with Notes on some European Institutions* (Washington: U.S. Government Printing Office, 1905).

10. *Sun*, July 26, 1896.

11. *Carpentry and Building*, February 1901.

12. *The Larkin Idea*, November 1906.

13. *New York Times*, July 20, 1904.

3. For Paper, Not People

1. *Los Angeles Times*, November 16, 1909.

2. *Ohio State Bar Association: Proceedings of the Thirty-Fifth Annual Session of the Association*, 1914.

4. *Coolth:* Everybody's Doing It

1. Margaret Ingels, *Willis Haviland Carrier: Father of Air Conditioning* (New York: Garden City Press, 1952), p. 67.

2. *New York Times*, July 29, 1929.

3. *New York Times*, May 28, 1936.

4. *Los Angeles Times*, October 28, 1928.

5. *B'nai B'rith Magazine*, January 1930.

6. William C. Allen, "Installation of a New Ventilation and Air Conditioning System for the Senate Chamber," printed notice to senators, quoted in *History of the United States Capitol: A Chronicle of Design, Construction, and Politics* (Washington: U.S. Government Printing Office, 2001).

5. Big Ideas. Bold Concepts. Bad Timing.

1. *New York Times*, May 27, 1931.

2. *San Antonio Evening News*, November 9, 1928.

3. *Los Angeles Times*, June 8, 1931.

6. From Home Front to Each Home

1. *Wall Street Journal*, August 18, 1944.

2. *Better Homes and Gardens*, February 1949.

7. The Unnecessary, Unhealthy Luxury (That No One Would Give Up)

1. *The New Yorker*, July 4, 1959.

2. *The Press and the Building of Cities: Proceedings of a Working Conference for 30 Reporters from Metropolitan Newspapers*, Columbia University/American Institute of Architects, 1962.

Bibliography

Books

Ackermann, Marsha E. *Cool Comfort: America's Romance with Air-Conditioning* (Washington: Smithsonian Institution Press, 2002).

Albrecht, Donald, and Chrysanthe B. Broikos. *On the Job: Design and the American Office (Part 3)* (New York: Princeton Architectural Press, 2000).

Allen, William C. *History of the United States Capitol: A Chronicle of Design, Construction, and Politics* (Washington: U.S. Government Printing Office, 2001).

Ansbro, George, and Leonard Maltin. *I Have a Lady in the Balcony: Memoirs of a Broadcaster in Radio and Television* (Jefferson, N.C.: McFarland, 2009).

Architect of the Capitol. *Report of the Architect of the United States Capitol, together with Report of Prof. S. H. Woodbridge relative to Improving Ventilation of Senate Wing of Capitol* (Washington: U.S. Government Printing Office, 1897).

Arnott, Neil. *On Warming and Ventilating: With Directions for Making and Using the Thermometer-stove, Or Self-regulating Fire, and Other New Apparatus* (Longman, Orme, Brown, Green, and Longmans, 1838).

Baldwin, Karl F. *History and Methods of Air Conditioning on the Baltimore and Ohio Railroad* (manuscript, 1934).

Banham, Reyner. *The Architecture of the Well-tempered Environment* (Chicago: University of Chicago Press, 1984).

Barnouw, Erik. *A History of Broadcasting in the United States, Volume 2: The Golden Web, 1933 to 1953* (New York: Oxford University Press, 1968).

———. *A Tower in Babel: A History of Broadcasting in the United States to 1933* (New York: Oxford University Press, 1966).

Becker, Raymond B. *John Gorrie, M.D.: Father of Air Conditioning and Mechanical Refrigeration* (New York: Carlton Press, 1972).

Benson, Susan Porter. *Counter Cultures: Saleswomen, Managers, and Customers in American Department Stores, 1890–1940* (Champaign: University of Illinois Press, 1987).

Bernan, Walter. *On the history and art of warming and ventilating rooms and buildings, by open fires, hypocausts, German, Dutch, Russian, and Swedish stoves, steam, hot water, heated air, heat of animals, and other methods; with notices of the progress of personal and fireside comfort, and of the management of fuel. Illustrated by two hundred and forty figures of apparatus* (George Bell, 1845).

Binder, Georges, *Taipei 101* (Mulgrave, Victoria: The Images Publishing Group, 2008).

Borsodi, William, ed. *Bakery and Confectionery Advertising: A Collection of Selling Phrases, Descriptions, and Illustrated Advertisements as Used by Successful Advertisers* (Advertisers' Cyclopedia Company, 1910).

Boughey, Davidson. *The Film Industry* (Sir Isaac Pitman & Sons, Ltd., 1921).

Broadbent, Geoffrey. *Emerging Concepts in Urban Space Design* (London: E & FN Spon, 1990).

Building Research Advisory Board. *Weather and the Building Industry: A Research Correlation Conference on Climatological Research and its Impact on Building Design, Construction, Materials and Equipment* (National Academy of Sciences, 1950).

Burrows, Edwin G., and Mike Wallace. *Gotham: A History of New York City to 1898* (New York: Oxford University Press, 1999).

Carrier Engineering Corporation. *The Story of Mechanical Weather: By the Mechanical Weather Man* (1919).

A Century of Progress International Exposition. *Official Guide Book of the World's Fair of 1934* (Chicago: The Cuneo Press, 1934).

Chartered Institution of Building Services Engineers. *The Story of Comfort Air Conditioning* (eBook), 2009.

Clough, Albert L. *A Dictionary of Automobile Terms* (The Horseless Age Company, 1913).

Cooper, Gail. *Air-Conditioning America* (Baltimore: Johns Hopkins University Press, 1998).

Cox, Stan. *Losing Our Cool: Uncomfortable Truths About Our Air-Conditioned World (and Finding New Ways to Get Through the Summer)* (New York: The New Press, 2010).

Cromley, Elizabeth Collins. *Alone Together: A History of New York's Early Apartments* (Ithaca, N.Y.: Cornell University Press, 1990).

Cudahy, Brian J. *A Century of Subways: Celebrating 100 Years of New York's Underground Railways* (New York: Fordham University Press, 2003).

Daniels, Rudolph L. *Trains Across the Continent: North American Railroad History* (Bloomington: Indiana University Press, 1997).

Delatour, Albert J. *A Daily Record of the Thermometer for Ten Years from 1840 to 1850* (R. Craighead, 1850).

Donaldson, Barry, and Bernard Nagengast. *Heat and Cold: Mastering the Great Indoors* (Atlanta: American Society of Heating, Refrigerating and Air-Conditioning Engineers, Inc., 1994).

Dupré, Judith. *Skyscrapers: A History of the World's Most Famous and Important Skyscrapers* (New York: Black Dog & Leventhal, 1996).

Edwards, Frederick, Jr. *The Ventilation of Dwelling Houses and the Utilization of Waste Heat From Open Fire-Places* (Longmans, Green & Co., 1881).

Emery, Fred. *Watergate: The Corruption of American Politics and the Fall of Richard Nixon* (New York: Times Books, 1994).

Erenberg, Lewis A. *Steppin' Out: New York Nightlife and the Transformation of American Culture, 1890–1930* (Chicago: University of Chicago Press, 1984).

Fischler, Stan. *Uptown, Downtown* (New York: Dutton, 1976).

Fisher, David E., and Marshall Jon Fisher. *Tube: The Invention of Television* (New York: Harcourt Brace & Company, 1996).

Fitzgerald, F. Scott. *The Great Gatsby* (New York: Charles Scribner's Sons, 1925).

Gerhard, William Paul. *Theatres: Their Safety from Fire and Panic, Their Comfort and Healthfulness* (Bates & Guild, 1900).

Gelman, Andrew. *Red State, Blue State, Rich State, Poor State: Why Americans Vote the Way They Do* (Princeton, N.J.: Princeton University Press, 2008).

Gomery, Douglas. *Shared Pleasures: A History of Movie Presentation in the United States* (Madison: University of Wisconsin Press, 1992).

Gouge, Henry Albert. *New System of Ventilation: Which Has Been Thoroughly Tested under the Patronage of Many Distinguished Persons* (Union Steam Presses, 1881).

Gugliotta, Guy. *Freedom's Cap: The United States Capitol and the Coming of the Civil War* (New York: Hill and Wang, 2012).

Hardy, Lady Duffus. *Through Cities and Prairie Lands: Sketches of an American Tour* (R. Worthington, 1881).

Hark, Ina Rae, ed. *Exhibition, the Film Reader* (New York: Routledge, 2002).

Hart, Moss. *Act One: An Autobiography* (New York: Random House, 1959).

Hartshorne, Henry. *Essentials of the Principles and Practice of Medicine* (Henry C. Lea's Son & Co., 1881).

Heard, J.A.E. (Archie). *Willis H. Carrier: The Man and His Message* (manuscript, 1979).

Hendrickson, Robert. *The Grand Emporiums: The Illustrated History of America's Great Department Stores* (New York: Stein and Day, 1980).

Herbert, Stephen, ed. *A History of Early Television, Volume 2* (New York: Routledge, 2004).

Hewett, Waterman Thomas. *Cornell University: A History (Volume Two)* (University Publishing Society, 1905).

Hood, Thomas, and Frances Freeling Broderip. *The Works of Thomas Hood* (E. Moxon, Son, & Co., 1873).

House of Commons Public Administration Select Committee. *House of Commons Accommodation: Third Report of Session 2005–06* (London: Stationery Office, 2006).

Hubbard, Elbert. *A Dozen and Two Pastelles in Prose: Being Impressions of the Wanna-maker Stores Written in as many Moods* (The Roycrofters, 1907).

Hughes, Rupert. *The Real New York* (The Smart Set Publishing Company, 1904).

Ierley, Merritt. *Comforts of Home: The American House and the Evolution of Modern Convenience* (New York: Three Rivers Press, 1999).

Imperial War Museum, *The Cabinet War Rooms* (1996).

Ingels, Margaret. *Willis Haviland Carrier: Father of Air Conditioning* (New York: Garden City Press, 1952).

Johnson, Steven. *Where Good Ideas Come From: The Natural History of Innovation* (New York: Penguin, 2010).

Knight, Edward Henry. *Knight's American Mechanical Dictionary: Being a Description of Tools, Instruments, Machines, Processes, and Engineering: History of Inventions: General Technological Vocabulary: and Digest of Mechanical Appliances in Science and the Arts, Volume 3* (Houghton Mifflin, 1876).

Kohn, Edward P. *Hot Time in the Old Town: The Great Heat Wave of 1896 and the Making of Theodore Roosevelt* (New York: Basic Books, 2010).

Kolodin, Irving. *The Metropolitan Opera, 1883–1966: A Candid History* (New York: Knopf, 1966).

Krupp'sche Gussstahlfabrik. *Exhibition Catalogue of the Cast Steel Works of Fried. Krupp, Essen on the Ruhr (Rhenish Prussia)* (Essen, 1893).

Landau, Sarah Bradford, and George Browne Post. *George B. Post, Architect: Picturesque Designer and Determined Realist* (New York: Monacelli Press, 1998).

Lewis, Logan. *The Romance of Air Conditioning* (Carrier Corporation, 1950).

Marchand, Roland. *Advertising the American Dream: Making Way for Modernity, 1920–1940* (Berkeley: University of California Press, 1985).

Matthews, Chris. *Jack Kennedy: Elusive Hero* (New York: Simon & Schuster, 2011).

Mayer, Dale C. *Lou Henry Hoover: A Prototype for First Ladies* (Hauppauge, N.Y.: Nova History Publications, 2004).

Mayer, Martin. *The Met: One Hundred Years of Grand Opera* (New York: Simon & Schuster, 1983).

Meigs, Montgomery C. *Capitol Builder: The Shorthand Journals of Montgomery C. Meigs, 1853–1859, 1861* (Washington: U.S. Government Printing Office, 2002).

Metz, Robert. *CBS: Reflections in a Bloodshot Eye* (Playboy Press, 1975).

Millett, Larry. *Once There Were Castles: Lost Mansions and Estates of the Twin Cities* (Minneapolis: University of Minnesota Press, 2011).

Moe, Kiel. *Thermally Active Surfaces in Architecture* (New York: Princeton Architectural Press, 2010).

Moyer, James Ambrose, and Raymond Underwood Fittz. *Air Conditioning* (New York: McGraw-Hill, 1933).

Nachman, Gerald. *Right Here on Our Stage Tonight!: Ed Sullivan's America* (Berkeley: University of California Press, 2009).

Navy Department. *Annual Report of the Chief of the Bureau of Steam Engineering for the Year 1881* (Washington: U.S. Government Printing Office, 1882).

———. *Reports of Officers of the Navy on Ventilating and Cooling the Executive Mansion During the Illness of President Garfield* (Washington: U.S. Government Printing Office, 1882).

Nebeker, Frederik. *Dawn of the Electronic Age: Electrical Technologies in the Shaping of the Modern World, 1919 to 1945* (Hoboken, N.J.: Wiley, 2009).

Packard, Vance. *The Hidden Persuaders* (New York: Penguin, 1957).

———. *The Status Seekers* (New York: Penguin, 1959).

Peters, Charles. *Five Days in Philadelphia: 1940, Wendell Willkie, FDR and the Political Convention That Won World War II* (New York: PublicAffairs, 2005).

Quinan, Jack. *Frank Lloyd Wright's Larkin Building: Myth and Fact* (Chicago: University of Chicago Press, 1987).

Reid, David Boswell. *Illustrations of the Theory and Practice of Ventilation* (Longman, Brown, Green, and Longmans, 1844).

Roberts, Brian M. *The Story of Comfort Air Conditioning* (Heating, Ventilating & Air Conditioning Manufacturers Association, 2009) (eBook).

Rome, Adam. *The Bulldozer in the Countryside: Suburban Sprawl and the Rise of American Environmentalism* (New York: Cambridge University Press, 2001).

St. Regis Hotel Company. *The St. Regis Hotel* (Grafton Press, 1905).

Sargent, Epes Winthrop. *Picture Theatre Advertising* (Chalmers Publishing Company, 1915).

Schroeder, Alan. *Presidential Debates: Fifty Years of High-Risk TV* (New York: Columbia University Press, 2008).

Schultz, Eric B. *Weathermakers to the World: The Story of a Company, The Standard of an Industry* (Carrier Corporation, 2012).

Seale, William. *The White House: The History of an American Idea* (Washington: White House Historical Association, 1992).

———. *The President's House: A History* (Washington: White House Historical Association, 2008).

Shepherd, Robert. *Westminster: A Biography: From Earliest Times to the Present* (New York: Bloomsbury, 2012).

Sherlock, Vivian M. *The Fever Man: A Biography of Dr. John Gorrie* (Aurora, Ill.: Medallion Press, 1982).

Sokalski, J. A. *Pictorial Illusionism: The Theatre of Steele MacKaye* (Montreal: McGill-Queen's University Press, 2007).

Soper, George A., Ph.D. *The Air and Ventilation of Subways* (Wiley, 1908).

Stedman, Thomas L., ed. *Twentieth Century Practice: An International Encyclopedia of Modern Medical Science, by Leading Authorities of Europe and America* (William Wood and Company, 1895).

Stern, Robert A. M., David Fishman, and Thomas Mellins. *New York 1960: Architecture and Urbanism Between the Second World War and the Bicentennial* (New York: Monacelli Press, 1997).

Stern, Robert A. M., Thomas Mellins, and David Fishman. *New York 1880: Architecture and Urbanism in the Gilded Age* (New York: Monacelli Press, 1999).

Stern, Robert A. M., Gregory F. Gilmartin, and Thomas Mellins. *New York 1930: Architecture and Urbanism Between the Two World Wars* (New York: Rizzoli, 2009).

Sylvester, Charles. *The Philosophy of Domestic Economy: As Exemplified in the Mode of Warming, Ventilating, Washing, Drying, & Cooking, and in Various Arrangements Contributing to the Comfort and Convenience of Domestic Life, Adopted in the Derbyshire General Infirmary* (H. Barnett, 1819).

Tafel, Edgar. *Years with Frank Lloyd Wright: Apprentice to Genius* (New York: McGraw-Hill, 1979).

Valentine, Maggie. *The Show Starts on the Sidewalk: An Architectural History of the Movie Theatre* (New Haven, Conn.: Yale University Press, 1994).

Van Antwerp, William Clarkson. *The Stock Exchange from Within* (Doubleday, Page & Co., 1913).

Walker W. *Useful Hints on Ventilation; Explanatory of its leading principles, and designed to facilitate their application to all kinds of buildings* (J. T. Parkes, 1850).

John Wanamaker Co. *New York City and the Wanamaker Store* (Underwood & Underwood, 1916).

Welling, David. *Cinema Houston: From Nickelodeon to Megaplex* (Austin: University of Texas Press, 2007).

Whitaker, Jan. *Service and Style: How the American Department Store Fashioned the Middle Class* (New York: St. Martin's Press, 2006).

Articles

Arsenault, Raymond. "The End of the Long, Hot Summer: The Air Conditioner and Southern Culture," in *The Journal of Southern History*, November 1984.

Bhatti, Mohinder S., Ph.D. "Riding in Comfort: Part II," in *ASHRAE Journal*, September 1999.

Bruegmann, Robert. "Central Heating and Forced Ventilation: Origins and Effects on Architectural Design," in *Journal of the Society of Architectural Historians*, Vol. 37, No. 3, October 1978.

Dudley, Dr. Chas. B. "The Ventilation of Passenger Cars on Railroads," in *Journal of the Franklin Institute*, Vol. 144, July 1897.

Friedman, Robert. "The Air-Conditioned Century," *American Heritage*, August/September 1984.

Gladstone, John. "John Gorrie, The Visionary," in *ASHRAE Journal*, December 1998.

Greater London Authority, "Sustainable Design and Construction: The London Plan Supplementary Guidance," 2006.

Maidment, G. G., and J. F. "Sustainable Cooling Schemes for the London Underground Railway Network." Missenden, School of Engineering Systems and Design, South Bank University, 2001.

Meigs, Capt. M. C., United States Corps of Engineers, "Notes on Acoustics and Venti-
    lation, with reference to the new Halls of Congress," in *The Civil Engineer and
    Architect's Journal*, May 1853.
Nagengast, Bernard. "A History of Comfort Cooling Using Ice," in *ASHRAE Journal*,
    February 1999.
———. "Early Twentieth Century Air-Conditioning Engineering," in *ASHRAE Journal*,
    March 1999.
———. "100 Years of Air Conditioning," in *ASHRAE Journal*, June 2002.
Van du Zee, Paul. "New York Roof Gardens," in *Godey's Magazine*, August 1894.

Sources
Architect and Engineer
Architects' and Builders' Magazine
Architectural Record
Carpentry and Building
Engineering Review
Engineering World
Harper's New Monthly Magazine
The Heating and Ventilating Magazine
Ice and Refrigeration
Journal of the American Society of Mechanical Engineers
Modern Mechanix
Popular Mechanics
Popular Science
Scientific American
Street Railway Journal
The Technical World
Transactions of the American Society of Heating and Ventilating Engineers

Index